# Space Enterprise

David A. Dietzler

Copyright-©-2021-by David A. Dietzler-All-Rights-Reserved

# Table of Contents

Preface .................................................................................................................... 4

Capitalism and the Road to Space ........................................................................ 5

The Case for Space Solar Power .......................................................................... 17

Mass Production, Reusability and Scavenging for CATS ( Cheap Access to Space) ... 24

Off-Earth Manufacturing ...................................................................................... 29

Interplanetary Trade ............................................................................................ 38

Breaking it Down: Process Summary .................................................................. 44

Laser Telecommunication .................................................................................... 47

Pulsar Navigation ................................................................................................ 51

Lunar Sports ........................................................................................................ 53

Safety and Space ................................................................................................. 55

Thoughts About Space Settlement Construction ................................................ 58

Collision Avoidance in Space ............................................................................... 64

Space Food .......................................................................................................... 67

Orbital Clean-up .................................................................................................. 71

Lunar Railroads ................................................................................................... 72

Miscellaneous Products in Space ........................................................................ 76

Space Liners & Centrifuges ................................................................................. 80

Space Corporations ............................................................................................. 91

Space Ports Etc. ................................................................................................... 99

Farming on the Moon .......................................................................................... 104

Preventing Pandemics in Space .......................................................................... 108

Off-Earth Mining: Speculations ........................................................................... 111

Making it in Space ............................................................................................... 114

Hope for Mars? .................................................................................................... 117

The Space Taxi ........................................................................................................... 121
ISRU Energy Storage ................................................................................................. 124
Fire and Ice ............................................................................................................... 128
Dwarfs and Giants .................................................................................................... 136
Diet for a Small Solar System ................................................................................... 139
References ............................................................................................................... 142

# Preface

This book is a follow up to my previous book <u>Mining the Moon: Bootstrapping Space Industry.</u> It continues the discussion regarding the "how-to" of space industry and settlement and ways to make money from it. Space is expensive and nobody is going to invest money in it if they can't expect any returns. Once the first steps of climbing out of Earth's gravity well and establishing industry on the Moon and in orbit are taken, space resources will be made available and the cost of building things in space will plummet. If we consider the vast resources of the asteroids and other bodies in the solar system and eventually the galaxy, the return on investment approaches infinity. A space faring civilization without limits to growth could emerge. Asteroid impactors that threaten the Earth could be diverted, but returning ice ages and super volcanoes could make civilization on Earth nearly impossible in the future. The best place for a growing technological civilization may be in outer space.

# Capitalism and the Road to Space

Free enterprise is already taking humanity on the road to a space faring civilization. Satellites, most of them owned by telecommunication companies, are an over three hundred billion dollar a year industry.[1] Companies like SpaceX and Blue Origin are competing with Boeing and Lockheed for rocket launch business. Blue Origin hasn't orbited any payloads yet, but SpaceX has with its partly reusable Falcon rockets. Healthy competition drives down prices and lower prices for access to LEO (Low Earth Orbit) are coming soon.

Manned space travel for exploration, tourism and settlement at affordable prices for the common person ignites the imagination. The key is the development of a low cost rocket or space plane that can transport humans into space for $100,000 or less. This might be achieved with the SpaceX Super Heavy or Starship. One hundred thousand dollars is still a lot of money, but many upper middle-class people could borrow the money and move into a smaller house. At this price, space travel still seems like a luxury for the elite. People working two jobs for minimum wage aren't going to be able to afford this; however, there will be jobs in space for everyone from janitors to spaceship pilots. Many space lovers will get jobs in space and spend more time in orbit or on the Moon than the wealthy tourists do. Some people with a pioneering spirit will sell everything and move into space permanently and become the first settlers on the new frontier.

We can be certain that taxpayers are not going to support a government run space travel service for wealthy tourists with the promise of low priced space travel in the future for ordinary citizens. If the government did start up a profit making corporation it might help balance the budget and pay down the national debt, but it would probably ruin private sector competition. It's more likely that corporations will conduct the space travel and development activities of the future and the government will take its cut in the form of numerous taxes and license fees. Government involvement will take the form of safety regulation enforcement and law enforcement in space. Big business-big government partnerships like the ones that built the railroads and various infrastructure systems might form to create infrastructure in space. Huge energy companies facing the reality of climate change and the dwindling demand for fossil fuels might invest heavily in space energy production.

A market for sub-orbital and orbital rocket or space plane rides will emerge within ten years if Virgin Galactic, SpaceX and Blue Origin succeed with their projects. Orbital hotels made from inflatable modules could be assembled and tourists could vacation in Earth orbit. Space energy promises to be a most lucrative industry. The construction of gigantic solar power satellites to supply cheap clean power to the Earth and the mining of helium 3 on the Moon to fuel advanced fusion reactors could be big business in coming decades. The key will be the establishment of a lunar base that uses on-site resources to bootstrap industry to a sufficiently large scale. The Moon base would supply materials launched with mass drivers to builders in space. It could also build the thousands of helium 3 mining machines needed to supply fusion fuel. The possibility of a government project sort of like a TVA of space always exists, but it will probably be better to leave the space business to the private sector.

A 5GWe SSPS (Space Solar Power Satellite) could earn revenues of over $2 billion a year selling wholesale electricity at five cents a kilowatt-hour. There would be no fuel costs and very little manpower would be needed to operate the satellite, so the potential for substantial profits exists. The satellite should last a long time in the rust and corrosion free vacuum of space. There would be no storms, lightning strikes, earthquakes or floods to damage it. There would be no interference with wildlife and indigenous people. It would be out of reach by terrorists, although ground rectennas would be vulnerable.

It's been stated that silicon solar panels are degraded 8 times faster than panels on Earth by radiation, mostly from cosmic rays and solar proton events, in space beyond the magnetosphere.[2] Apparently, this figure is based on an average degradation rate of 0.25% per year on Earth and a 2% per year degradation rate for silicon panels, the most common and lowest cost technology, in space.[3] This won't be a problem for settlements in LEO but it will be troublesome for SSPSs in high orbit and stations beyond. It will be necessary to build solar thermal satellites or replace the solar panels every ten years or so. The satellite could be overbuilt with 20%

to 25% more silicon solar panels than the powersat is rated at, but this will only extend the service interval by a decade or so.

If the SSPS consists mostly of reflectors that concentrate sunlight onto a smaller photovoltaic module it might be easier and cheaper to replace the solar panels.  One problem with this is that silicon solar cells become less efficient at high temperatures. There is always the possibility of discovering photovoltaic materials that are not adversely effected by space radiation and high temperatures.  Lunar ilmenite, perovskite and multi-junction GaAs/Ge/InGaP solar cells that have worked under light as concentrated as 2,000 suns are candidates for use in SSPSs instead of silicon.[4]  The microwave generators might have a limited lifetime too. Even on Earth, powerplants have to be shut down and serviced periodically. Turbines must be rebalanced and nuclear reactors shut down for about a month every year for refueling. The costs must simply be absorbed.  For the most part, powersats will not require a lot of maintenance, and when they do the work can be done by human crews in space constructing newer satellites or by robots teleoperated by humans in ground stations.  The frame, reflectors and kilometer wide dish antenna will probably last for many decades, maybe a century. The ROI (Return On Investment) should be generous.

With materials from the Moon it should be possible to produce rocket propellant in space and do it far less expensively than by rocketing it up from Earth.  Lunar soil, or regolith, is 40% oxygen.  A hydrogen/oxygen rocket will use a propellant mixture containing 6/7 to 8/9 liquid oxygen. The Moon could supply the bulk of propellant used by spaceships traveling to high Earth orbit, the Moon or Mars.  The SSPS builders will smelt regolith to get metals like aluminum, magnesium and titanium.  Large quantities of by-product oxygen will result.

Powersat builders might have more silicon, iron and calcium than they can make use of.  Excess accumulations of powdered silicon and powdered metals will burn in pure oxygen and could be used as rocket fuel.  This could be sold to companies operating rocket ships in space.  Hydrogen could be extracted from lunar polar ices.  It could be combined with abundant silicon to make silane liquefied at minus 111.5 degrees Celsius. A silane/oxygen rocket will use more propellant overall than a hydrogen/oxygen rocket but it will use only half as much hydrogen.  It would use far more propellant than a nuclear thermal rocket using $LH_2$ but only one fifth as much hydrogen.[5] Since hydrogen is in short supply it would make sense to burn silane in space. Silane could also be mixed with powdered silicon which burns with 13,000 BTU/lb. and powdered metals to make a slurry fuel and stretch hydrogen supplies even further. Hydrogen could also be "piggy backed" into space with rockets carrying passengers or cargo. Rockets that can orbit 100 tons are foreseeable. One hundred passengers would only weigh about ten metric tons. Another 20 or so metric tons of $LH_2$ could easily be carried into space. That's enough to make 160 tons of silane which would be burned with 320 tons of LOX produced in space from regolith launched from the Moon. If we dare to be less conservative, a rocket that can move 100 tons to LEO with ten tons of passengers in a 20 ton ship, could also carry 70 tons of $LH_2$. That's enough to make 560 tons of silane. This could be burned with 1120 tons of LOX for a total of 1680 tons of propellant and that would be enough to send a fairly big ship to lunar orbit and back to LEO. Silane would be produced in LEO from lunar silicon and hydrogen from Earth. The ship would load up with enough silane for a round trip but only enough LOX to reach the Moon. At the Moon, the ship would load up on enough LOX for return. This would be very efficient, even with retro-rocketing into LEO instead of dangerous aerobraking and atmospheric re-entry at about 7 miles per second which would necessitate a massive heat shield that might have to be replaced after each flight.

With hydrogen production from polar ice on the Moon it would be possible to make silane and LOX and ship them to space stations in low Lunar orbit (LLO) or at Lagrange point 1. In this case, the inter-orbital lunar ships would not have to carry extra silane fuel for return flight and this would make it possible to use less propellant. When things cost so many dollars per kilogram it always helps to save weight and cargo mass.

It may be argued that billions of tons of ice exist in permanently shadowed lunar polar craters; so why worry about hydrogen? The reality is or will be that while the hydrogen is there, getting it won't be easy or cheap. Mining machines will have to work at temperatures just a few degrees above absolute zero and solar energy will be blocked by crater rims. The robotic miners will need nuclear power, beamed energy, miles long tethers/power cords or a lot of batteries.

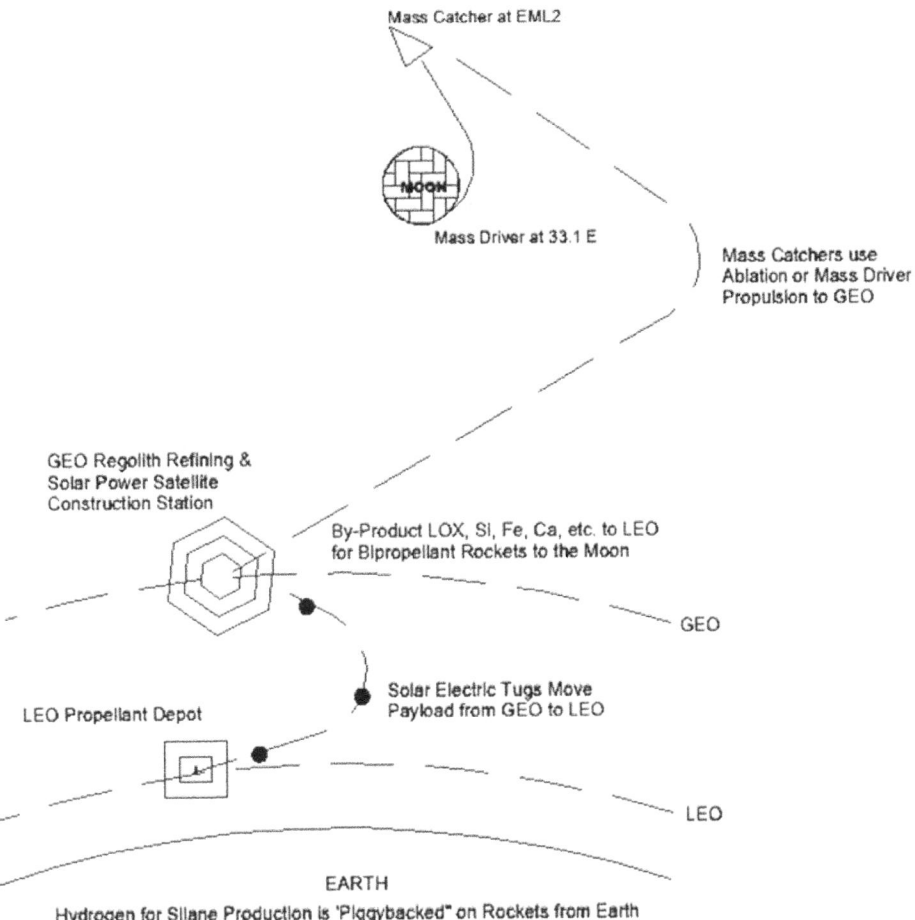

Fig. 1 Materials from the Moon supply SSPS construction in GEO and by-products are sold to LEO settlement builders and space rocket transportation companies.


With industry growing on the Moon it would become possible to build hotels and resorts. Vehicles for sight seeing trips could also be built. Railroads and high-speed trains will emerge thereby making it possible to visit many scenic locations on the Moon. With cheap propellant from the Moon lunar tourism could become a reality. Passengers would reach LEO via rocket or space plane then transfer to a ship docked at a propellant depot. The ship will leave Earth orbit and travel to a station at Earth-Moon Lagrange point one or LLO (Low Lunar Orbit). There they will transfer to landers and descend to the surface of the Moon. Lunar tourism along with orbital tourism will probably be multi-billion dollar a year industries someday.

Silicon dioxide is the main component of glass and there will probably be more of this than is needed to make solar panels in space. Glass has many structural uses. Iron is too heavy for spaceships and probably too heavy for SSPSs. It could be combined with carbon in solar heated crucibles to make steel. Steel might seem inferior to aluminum and titanium but it is a very useful well understood material that can be used to build lightweight structures due to its high strength to weight ratio. A small amount of carbon can make a large quantity of steel. The carbon could come from the Moon or even be "piggy backed" up from Earth. Calcium oxide and plain regolith can be used to make concrete. This cannot be poured in the vacuum of space but it could be used inside pressurized space stations for all sorts of constructions. With some cement, regolith, aluminum powder, water and a large autoclave it would be possible to make a wonderful strong lightweight material called AAC (Autoclaved Aereated Concrete). It could be used for walls, roofs, doors, furniture and more. Plain regolith could serve as a growth medium for plants of all sorts.

Steel and glass could be used to build LEO space settlements like Kalpana Two. These could serve as orbital hotels or contain time shares and private condos. Orbital real estate and tourism enterprises might be seen in the future. These settlements would rotate fast enough to produce 1G. In decks closer to the hub the force will be lower and in the hub it will be zero. Tourists could get a taste of low gravity and complete weightlessness in these stations. They could enjoy fantastic views of Earth below.

This kind of space settlement at 250 meters in diameter would be smaller than the Bernal Spheres and Stanford Toruses envisioned in the 1970s but they would be much easier to build. In Earth orbit they would be protected from cosmic rays by Earth's geomagnetic field, so they wouldn't need incredibly massive radiation shields and subsequently massive metal structures. They would be much closer to home than settlements stationed at Lagrange points.[6] It might take only ten minutes to rocket up to a station in ELEO (Equatorial Low Earth Orbit). In an emergency, quick return to Earth and landing at an equatorial base would be an option.

Images on next page. Copyright-©-Bryan Versteeg/Spacehabs.com
Used by permission.

Above: Exterior of Kalpana Two   Below: Interior of Kalpana Two

Given the high cost of space transportation it wouldn't make financial sense to send space workers and spaceship crews back to Earth when they were off duty or on the weekends and holidays.  They could stay in one of these

Kalpana settlements. Multi-level farms could also be built inside to provide food for tourists, condo and time share owners, workers, spaceship crews and settlers bound for Mars.

Space energy could be the driver that makes lunar industry and space industry profitable. It could bring space resources within reach. Metals and propellants from the Moon could also be applied to asteroid mining. Spacecraft for mining asteroids could be built in orbit and fueled with lunar materials.

There are about 20,000 known NEOs (Near Earth Objects). Many of these contain iron, nickel and cobalt along with gold, silver and platinum group metals. Others contain significant amounts of water that can be used for life support or as a source of hydrogen. Carbon, phosphorus and ammonia are present in some asteroids. These could be used as fertilizer in space farms. Hydrogen, carbon and nitrogen from ammonia could be used to make plastics. Asteroids also contain iron, magnesium, silicon and oxygen in the form of iron and magnesium silicates along with aluminum and calcium. All of these elements could be used in space for construction, life support or propellant. The first trillionaire might be the owner of an asteroid mining company.

Mars settlement and terraforming become possible with lunar and asteroid materials and space industry. Ships could be built on Earth and launched to LEO where they fuel up with propellant made in space and hit the trail to Mars. Much of the equipment Mars settlers will need to survive and prosper on Mars could be made on the Moon or in space manufacturing facilities. The settlers will spend a lot of their money to buy transport to LEO, spaceships, propellant, industrial equipment, farm supplies and everything else they will need. They will cut the cord with Earth and form completely independent communities on the red planet. They will form their own internal economies. Space industrialists will make money selling things to the settlers, but what will the Martians have to offer in return? It seems they would want to buy telecommunication services from Earth. Could Mars settlers sell things like copper or chlorine, present on Mars but almost non-existent on the Moon, to the Lunans? Will tourists travel to Mars? A trade triangle might form between Earth, the Moon and Mars. Interplanetary commerce could employ large populations in space.

Ultimately, Mars will be terraformed.  The descendants of the early Martian settlers might lay claim to vast territories on Mars and sell land to those who come later to farm, mine and live. They might back-up their territorial claims with armed force.  This might offend those with idealistic notions of space resources shared by all of humanity, but private property is a fact of life and one of the foundations of happiness.

Mars may evoke a lot of passion in enthusiastic would be settlers, but asteroids might be where the money is.  Most asteroids are C-type and many of them contain hydrogen, carbon and nitrogen bearing organic chemicals in the form of a tarry substance resembling kerogen.  They also contain 3% to 22% water based on the study of meteorites.  These materials will be of great value in space.  They could provide rocket propellant, fertilizer, plastics, synthetic fibers, chemicals, paint, dye, drugs, lubricants and more.  Oil is our primary source of organic chemicals on Earth that are used to make thousands of different products. In space, settlers on the Moon and in free space habitat will look to the asteroids for these things.  With lunar materials for spacecraft construction and propellant it will become possible for robotic spaceships with AI electronic brains to mine NEOs.  The delta velocity to reach these bodies is not very high but launch windows to and from these asteroids will be years apart.  Asteroid mining is a task best left to robots the require minimal human supervision by radio or laser communication links.  The robots might mine these asteroids and just haul payloads of organics and water back to Earth-Moon space or they might mine the asteroid and use that matter for mass driver propellant to push the whole asteroid into orbit around the Earth or Moon.

There are S-type asteroids that consist of rock and metal and M-type asteroids that are entirely metal mostly in the form of iron, nickel and cobalt.  With carbon from C-type asteroids this metal could be converted into high strength steel for construction of free space habitat miles in diameter. Metallic asteroids also contain precious metals including gold, silver and platinum group metals (PGMs).  While C-type asteroids consist of material that is not too hard to mine, the M-type asteroids could present a problem.  High energy lasers, electron beams or plasma cutting jets might be needed to cut up the metal.  Devices similar to huge mass spectrometers might be used to separate the metals so that alloy steels could be made and precious metals could be extracted in pure form.

Many have envisioned a solar system with millions of free space settlements in addition to planets, moons and asteroids with perhaps a trillion or more people. The demand for water and organic chemicals will be immense. Beyond asteroids, there are the carbon dioxide rich clouds of Venus, the HCN rich atmospheres of the Gas Giants and Ice Giants, the ices of many planetary moons and the atmosphere and methane lakes of Saturn's largest moon Titan. What if heavier hydrocarbons formed on Titan and sank to the bottom of its methane lakes? That would be like striking oil in space but this is pure speculation. We do know that Titan's atmosphere contains nitrogen, methane and ethane. What if it proved to be economical and lucrative to pump up liquid methane from the lakes on Titan and rocket it into space for use by settlers? Very efficient nuclear thermal rockets using liquid methane as working fluid could be used to transport the stuff. Nitrogen in Titan's atmosphere could be combined with hydrogen from methane and ethane to make ammonia which would be much easier to store and transport. It's not possible to predict a business case for these various sources of light elements, but we can be certain that vast resources exist in our solar system. While the early years of industry on the Moon will struggle with the dearth of light elements and water, this will not be the case in space forever.

The atmospheres of Jupiter, Saturn, Uranus and Neptune contain vast amounts of helium including the potential fusion fuel helium 3. Normal helium 4 has two protons and two neutrons. Helium 3 has two protons and one neutron. The fusion of deuterium and tritium releases floods of high energy neutrons that could damage the reactor and shorten its lifetime. Deuterium fused with deuterium also generates neutrons. Helium 3 and deuterium produces an alpha particle and a proton but some neutrons are generated by deuterium-deuterium side reactions in the plasma. Helium 3 fused with helium 3 produces no neutrons. If it ever becomes possible to fuse helium 3 with lasers or tokamaks with high temperature superconducting magnets it would be the ideal fusion fuel and there's enough of it in the atmospheres of the outer planets to power human civilization for millions of years. It's not possible to predict a business case for helium 3 at the present time, but if we ever develop the technology needed to make use of it there's plenty available.  Balloon borne robotic factories floating in the atmosphere of Saturn or huge ram-jets scooping up gases to obtain HCN and helium have been envisioned.  It's hard to imagine ever depleting all those resources, even with a trillion consumers and solar system wide trade networks.

A trillion human beings appeals to more than just people who like to make babies.  It appeals to people with an eye on the money to be made catering to the needs and wants of a thousand times as many people as there are on Earth today with living standards enjoyed by only the wealthiest among us.  Corporations that dwarf the likes of Exxon-Mobile, British Petroleum, Amazon or Wal-Mart could exist someday.  Grunt labor might be done entirely by robots while humans engage in intellectual and creative work.  There's no telling what music or movies will be like in the distant future; however, we can be certain that there will be solar system wide entertainment markets.  A hundred billion fans would definitely enrich pop-stars.

Trillionaires, deca-trillionaires and cento-trillionaires could appear.  Ordinary people might live merely like today's millionaires.  Extended life spans due to advances in medical science and genetics would make it possible to amass more wealth over the years, decades, even centuries.  Thousands of robotic freighters and tankers would traverse the solar system carrying raw materials and finished products.  People would travel on luxurious space cruise liners.  Vast numbers of people could live in free space settlements where the weather is always mild and visit many if not most of the worlds of the solar system if we don't count the millions of asteroids.  Life in a settlement will not be like life in a zoo.  In a habitat like Island 3 the outdoors would be much like that on the surface of a planet and with the exception of some occasional gentle rains the weather would always be ideal.  Those who long for mountains will travel to Earth, the Moon, Mars and other scenic worlds.  Millions of people today live in cities and get out into the country or travel to the mountains and sea shores during their time off work and they are perfectly happy.  Life in space will not be confinement to a spinning metal cylinder with as much joy as a prisoner.  It will be very exotic.

Dystopian nightmares and socialist clone societies are not to be expected. Work will still be expected and jobs done by humans will require lots of education or lots of talent or both. Mercantile activities, money and personal possessions will not disappear in some kind of cult like existence for swarms of genetically engineered minions. Rewards will be earned and there will be freedom from need and want even though some may have more than others. Democracy and free markets will prevail.

# The Case for Space Solar Power

A solar panel on the ground will be illuminated 29% of the time on the average. A solar panel in GEO will be illuminated 100% of the time. That's 3.45 times as much. Sunlight is also 144% more intense in space, but much of that is infrared light.[7] To tap all that energy a solar cell made of something other than silicon may be needed. If solar cells that convert the full spectrum of sunlight to energy can be applied, a solar panel in space could produce five times as much energy as one on the ground.

Microwave transmission of power is about 85% efficient and conversion of electricity to microwaves is 60% to 95% efficient.[8] Therefore, a solar panel in space can deliver 2.74 to 3.82 times as much energy to the ground via microwave beam than a ground based solar panel can under the best conditions.

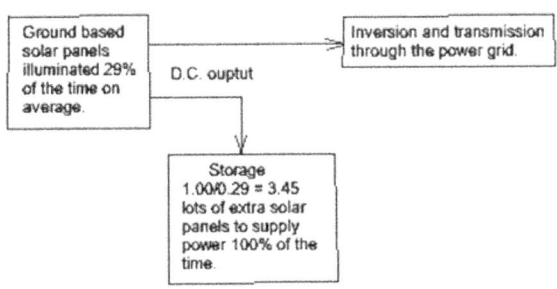

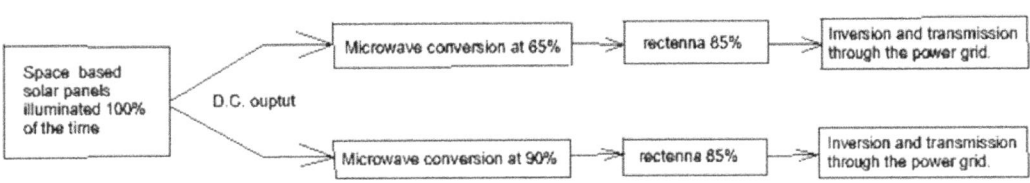

Space based solar panels collect 1.00/0.29 = 3.45x energy as same area of solar panels on Earth.

Solar radiation in space outside Earth's atmosphere is 144% more intense.

3.45 x 1.44 = 5x  Solar panels in space collect 5x as much energy as equal area on Earth.

5 x 0.65 x 0.85 = 2.74x as much electricity as ground based solar even after conversions.

5 x 0.9 x 0.85 = 3.82x as much as ground based even after conversions.

Atmospheric absorption by water vapor in clouds could weaken the beam, but this would not always be a problem. At microwave frequencies that are transmitted by the atmosphere beam divergence is a challenge. It becomes necessary to use a large aperture dish antenna that is perhaps a kilometer in diameter and a large receiving antenna (rectenna) with a diameter of ten kilometers or more. The challenge of building a large rectenna is not that great. Huge dish antennas in space might be made by depositing aluminum on a giant inflatable template of some sort. Once the electricity is received on the ground it must be inverted from D.C. to A.C. and there will be some loss of power, but conventional ground based solar faces the same limitation so there's no need for comparison here.

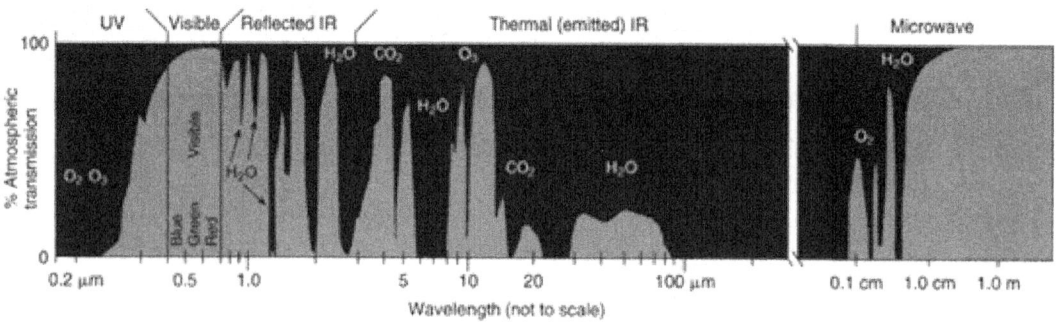

This energy is available 24/7 in all kinds of weather and fuel costs for its production are non-existent. There is no need for expensive energy storage systems to supply power at night, when the skies are overcast or the wind doesn't blow. Space solar power may be cost competitive with ground based renewable energy systems. Solar power satellites are said to lack load following capability. This problem also exists for ground based solar and wind power. Might it be possible to use SSPSs for baseload power and kick in dormant satellites during intermediate and peak power demand time periods? Could other ground based systems like molten salt reactors, natural gas fired turbines, batteries, flywheels, pumped hydroelectric storage, conventional hydropower and underground compressed air storage be used to make load following possible? What about diverting excess satellite power to areas where sunshine and wind are lacking at times through a nation-wide or continental power grid controlled by computers with some long distance superconducting connections?

The environmental footprint of space solar power is very small. Mountain tops don't have to be removed for coal. Deep rock formations don't have to be fractured. Radioactive wastes don't have to be transported or stored. No carbon dioxide, sulfur or nitrogen oxides are emitted. Thousands of acres don't have to be covered with solar panels. Crops can still be cultivated underneath rectennas.

Implementation of space solar power will depend on costs; especially the cost of rocketing cargo into space. Since the cost of climbing out of Earth's gravity well to LEO is high, it seems using lunar materials would be more economical. The next challenge would be building SSPSs with those materials. Magnesium could be hot rolled into sheets to make reflectors and aluminum could be rolled into sheet metal and plates that are cut up into strips and shaped to make frame members for the satellite.

In 1978 researchers at Grumman built a "Space Fabrication Demonstration System". This device is also called a beam-builder. It can build triangular beams from rolls of aluminum sheet metal and cross braces that are made separately. The machine shapes long strips of aluminum into angled pieces and welds them to the cross braces. A one-hundred foot beam weighs just 85 pounds but can support a load of 1,260 pounds. The beam builder can make a thousand foot long beam in just two hours.[9] This is the kind of automated technology needed for solar power satellite construction. A powersat will consist mostly of beams that form a framework to support reflectors that concentrate solar energy onto a photovoltaic module.

Fig. 2 A beam-builder at work in space. NASA

The beam-builder would weigh 23,272 pounds (10,587 kg.).[10] Due to its complexity, it would be made on Earth and shipped into space with a commercial rocket. Space construction stations built in GEO from parts rocketed up to LEO then hauled with solar-electric tugs to GEO would extract aluminum and other metals from lunar regolith. Aluminum could be alloyed with lunar manganese, silicon and/or magnesium. These would not be the best alloys but they would have to do. Copper, lithium and zinc for making the best aluminum alloys are not present in any significant amounts on the Moon. Metals would be extracted by a combination of electrical and chemical methods, melted and cast into slabs. The slabs would then be rolled into sheet metal for the beam builders.

Rolling mills are massive pieces of equipment, but they are probably faster than any kind of electrothermal spraying in the vacuum. Rather than launch the rolling mills to GEO, they would be constructed at the space stations by extracting iron from regolith, converting it to steel in crucibles, sand casting the steel in sand molds bound with sodium silicate to make the rolls and heavy frame parts, then CNC machining of the castings. Smaller parts for the rolling mills could be made by 3D printing and some parts could be rocketed up from Earth. The rolling mills would crank out aluminum sheet and other products like foils, plates, bars, heavy I or H beams, rods, rails and pipes. All this would have a mass far less than the tens of thousands of tons of metal parts needed for powersats.

The construction stations would consist of aluminum girders welded, riveted or bolted together. Some would rotate to provide a low amount of "artificial gravity" so that molten metals could be poured and other jobs could be done in a more natural way. Others would be weightless work platforms. Once the first station is assembled and in operation it can make parts for more stations that are assembled in orbit by teleoperated machines and autonomous robots. The first company to put a station in action could sell its services to other companies that want a station to assemble numerous solar power satellites of their own. These powersats would be sold to utility companies or they would just sell wholesale electricity to utilities on the ground that build their own rectennas.

At two cents per kilowatt hour a 5GWe satellite could earn revenues of $876 million per year. At five cents per kilowatt hour; $2.19 billion per year.

Maintenance and operating costs would be low. This carbon and radioactive waste free electricity could be used to make hydrogen from water by electrolysis, desalinate sea water for irrigation programs that reclaim marginal and desert lands, power industry and promote the development of poor nations.

Space solar power could make a significant contribution to the future energy production mix. Fossil fuel companies might be wise to stop strip mining and well drilling and put money into space industrialization. Lots of infrastructure must come first from bootstrapping Moon bases and mass driver lunar launchers to GEO construction stations and beam builders. Space X Falcon rockets are working now. The Super Heavy and Blue Origin rockets have yet to be demonstrated. Even so, the job could be started today with Falcon rockets. If lower launch costs are realized in the future, the prospects of success are even greater.

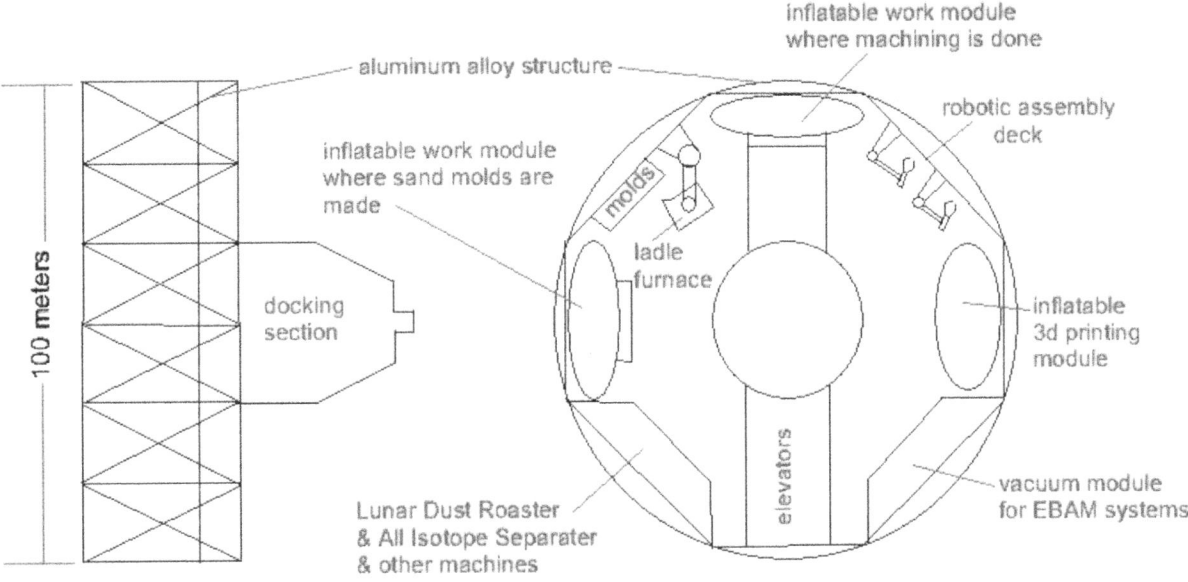

Fig. 3 Space station for making heavy equipment for SSPS construction.

| Years | Number of 5GWe SSPSs | Annual Revenue in Billions |
|---|---|---|
| 0 | 4 under construction | ---- |
| 1 | 4 | 3.5 |
| 2 | 8 | 7.01 |
| 3 | 12 | 10.51 |
| 4 | 16 | 14.02 |
| 5 | 20 | 17.52 |
| 5 YEAR SUBTOTAL | | 52.56 |
| 6 | 24 | 21.02 |
| 6 YEAR SUBTOTAL | | 73.58 |
| 7 | 28 | 24.53 |
| 7 YEAR SUBTOTAL | | 98.11 |

Fig. 4 At just 2 cents/kWhr. and construction of just 4 SSPSs per year, substantial revenues before costs can be accrued in only 7 years.

## Space Based Solar Energy Pros and Cons

### Pro

higher solar intensity, constant availability, never obscured by clouds, no night in GEO

is not intermittent

no fuel costs

no costs for energy storage

no costs for waste disposal, fuel reprocessing or plant decommissioning

no rust, corrosion, storms, lightning strikes, earthquakes, etc.

high capacity factor

potential for low cost electricity

no air and water pollution

no toxic or radioactive waste

no devastation of land

no transport by pipeline, truck, rail or ship

no dangers of explosion or spills

no displacement of wildlife or local populations

rectennas can be located anywhere

not effected by seasonal changes

no greenhouse gas emissions

easily integrated into existing power grids

## **Con**

large up front cost

no experience building super large structures in outer space

dangers for space workers

occasional meteor strikes

PV modules might need replacement every 20 years or so

# Mass Production, Reusability and Scavenging for CATS ( Cheap Access to Space)

**Mass Production**

Mass production is the key to making anything cheap. Automobiles, trucks, electronics, war materials and every other other product of our industrial age have become affordable thanks to assembly lines. The growing role of automation promises to make everything even less expensive. If we cranked out rockets on assembly lines they would be cheaper. The problem is that the market for all those rockets does not exist. A limited number of satellites are launched every year. If we struck up mass production of rockets like bombers for a war effort the result would be an enormous pile of hardware that could not be sold! High supply and low demand would spell bankruptcy for the rocket companies after they liquidated their stock. This is a case where the reverse law of supply and demand operates.

If there was a higher demand, supply could meet that demand for a reasonable price. However, as we all know, expendable rockets will never make any more fiscal sense than using a jet airliner once and then throwing it away. If space tourism is going to be the driving market force that leads to the opening of the high frontier and a space faring civilization then we must have reusable space vehicles.

It would also make sense to locate spacecraft assembly plants in better locations based on economics rather than politics. Launch procedures, recovery and maintenance might also be streamlined and more highly automated to reduce labor costs.  Equatorial launch bases that can take advantage of Earth's rotational velocity might also be built.  This could increase payload mass and reduce cost to LEO, as well as create jobs at the launch bases and nearby hotels for foreigners.  Private industry doesn't have to launch from American soil for national security reasons, but equatorial launch bases might necessitate spacecraft factories in other nations and American labor organizations might not be happy about that.

**Reusability**

Mass production will not work if we don't reuse launchers. Many of us dream of a single stage to orbit vehicle like the British Skylon. There are also high hopes for the SpaceX Starship which could orbit 100 passengers. When such vehicles are perfected and proven to be safe and reliable, the price to LEO will fall within a range that can be afforded by the millionaire class and professionals. A large global market will emerge. Almost everyone is interested in space tourism. Spaceplanes could be churned out in factories and spaceports with fueling facilities and perhaps mag-lev launching tracks will be built. At first the wealthy will take a ride into space but eventually space travel will drop in price until ordinary middle class citizens can afford an orbital jaunt.

Things won't stop there. A few hours in orbit will not be enough for many tourists. They will want prolonged stays in space and even travel to the Moon and Mars. This will require space stations in LEO, GEO, EML1, spacecraft that can reach other worlds and infrastructure to support those spacecraft. The Skylon will put 15 metric tons in LEO. That's enough payload capacity to move plenty of passengers and some inflatables for space stations.

A reusable heavy cargo lifter would help. A rocket based on a modified Shuttle external tank with LH2/LOX burning aerospike engines on the bottom and a cargo module on top along with a kerosene or methane and LOX powered sea barge landing booster that could put 100 to 200 tons of cargo in LEO would be ideal. The booster would be reusable and so would the hydrogen and oxygen burning engines. These engines would be part of a module with heat shield and parachutes that disconnects from the ET in space and returns to Earth. The ET and cargo module would be used in space. They would contain valuable titanium and aluminum alloys bearing copper and lithium, plastics, composites and fiberglass. Repurposing or recycling the ET and cargo module as well as finding uses for all cargo packaging materials will mean that they belong to the payload sent into space. Nothing would be wasted or thrown away. The relatively simple external tanks and cargo modules, compared to engines, could be mass produced on robotic assembly lines to cut costs.

Such a rocket could also be fitted with a large space capsule that can carry 50 to 100 passengers. Naturally, the capsule would be refurbished, refueled, and reused after every flight. It would parachute down for landing in a wide open area and fire retro rockets for a soft touchdown. If the retro rockets fail the heat shield could be ejected and an air bag could be inflated to cushion landing. Heat shields could be disposable. This could be cheaper than the high maintenance heat shield tiles of the Shuttle.

**Scavenging**

External tanks could be used for space station modules, spaceship hulls, propellant depot tanks, liquid storage tanks for space freighters that haul $NH_3$, $CH_4$, $H_2O$, organic chemicals, and scrap metal for space construction. The titanium alloys will contain aluminum and vanadium which is rare on the Moon. The aluminum alloys will contain copper and lithium which are also rare on the Moon. Since these alloys cannot be made with lunar or asteroidal materials this scrap metal will be very valuable. Pure aluminum from the Moon will not be good for much besides electrical wiring. Unalloyed aluminum is not very strong or hard. It could be alloyed with lunar magnesium, silicon, chromium and/or manganese to make half-way decent alloys but this will not suffice for high stress applications that demand copper and lithium aluminum alloys like the ones ETs are made of. For some applications ET scrap metal will be indispensable. Lunar titanium might be used instead of aluminum when high strength and light weight are required.

Cargo modules and cargo packaging materials consisting of plastics, composites, fiberglass, styrofoam and even cardboard will all be valued for their light elements--hydrogen, carbon and nitrogen. The modules might simply be cut up and the pieces machined into various products or melted down and injected into molds. Styrofoam and cardboard boxes might be valuable as is. They could also be decomposed with solar heat or plasma arcs to get light elements for Closed Ecological Life Support Systems.

Huge space hotels made from external tanks, inflatables and recycled materials will emerge. There will be no shortage of customers even when the millionaire class is the only group that can afford to stay in a space hotel. Global standards of living are rising. Barring an environmental catastrophe or a terrible world war there will be more millionaires than ever in the future and the middle class will be able to afford things that only rich people can have today. Billionaires, deca-billionaires and cento-billionaires, even trillionaires of the future will be able to afford their own space programs entirely! It's rather obvious that the tech giants of today will have the money to capitalize space travel industries of the future and this will not be a shameful luxury when it means the creation of jobs on Earth, jobs in space, and the birth of a space faring civilization for a world where frontiers have all but disappeared.

## Beyond LEO

Going beyond LEO requires large amounts of rocket propellant. The only way to get it for a reasonable price will be to bootstrap up industrial bases on the Moon. Polar ice and lunar regolith could supply hydrogen, oxygen and metal powders including silicon which could be combined with hydrogen to make silane, $SiH_4$, to fuel inter-lunar spacecraft. Eventually asteroids could be mined for propellant but common sense dictates that the development of the Moon should come first.

Inter-lunar spacecraft will include tugs with solar electric propulsion that use only small masses of propellant hauled up from Earth. These tugs will move cargos from LEO to a station at EML1. They will be produced in factories on Earth, rocketed to LEO, and reused. When their service lifetime ends they will be dismantled for the gallium, arsenic, indium, germanium, phosphorus and selenium in their solar panels to make new solar panels and the aluminum alloys, plastics and composites that compose their structures will be repurposed. Landers or "Moon Shuttles" that move cargo from EML1 to the lunar surface where they deposit their cargo, are refueled, and ascend back to EML1 numerous times will meet a similar fate.

Mass production and reusable hardware will make space travel affordable, but even then it will be pricy. In orbital space there is nothing but free vacuum, lots of solar energy, and microgravity that can be useful for containerless manufacturing and electrophoresis. There is quite a bit of space junk in Earth orbit and this may or may not be worth some money to salvage. External tanks, cargo modules and packaging that contain elements rare on the Moon and therefore unavailable in space even after the Moon is industrialized will be valued commodities.

If a time comes when hundreds, even thousands, of cargo rockets are leaving Earth every year to support a flourishing space hotel industry and maybe a microgravity manufacturing industry, then there should be plenty of titanium and aluminum alloys that can't be made on the Moon and other materials rich in hydrogen, carbon and nitrogen available in space. Space dwellers will learn to use the resources of space like lunar and martian basalt, low and mid-grade titanium and aluminum alloys, iron, magnesium, stone, glass, and platinum group metals from lunar meteoric iron-nickel particles and Near Earth Asteroids.

We must wonder about industrial processes and alloys that could use platinum group metals (PGMs) but are not used commonly due to their high cost. When asteroidal PGMs become abundant this might change. On Earth, copper, lithium, tin, vanadium, zinc, thorium, zirconium, yttrium and many other metals are essential for industry. What if PGMs can substitute for or even surpass these metals in different applications? The Spaniards found a curious "white silver" later determined to be platinum possessed by the South American Indians that made their cannons stronger. Could steel alloys made with PGMs that are uneconomical today but might be cost effective in the future when asteroids are mined be possible? Even then, in the deserts of outer space, the Moon, Mars and beyond everything will be reused and recycled, especially if it came from a deep gravity well.

# Off-Earth Manufacturing

On Earth, mass production has made an abundance of goods cheap enough to be afforded by the bulk of the population. Mines and refineries produce raw materials like steel, aluminum and petrochemicals. Enormous numbers of factories receive these materials and each produces a few parts or just a single part. The parts are all shipped to more factories where they are assembled into a single product. For example, one factory makes carburetors (or EFI systems these days), another makes tires, yet another makes transmissions and others make more car components. The parts all come together at an assembly plant and a car or truck is made. Thousands of specialized factories each rapidly producing large numbers of copies of just one or a few products work together to make modern industrial civilization possible.

On the Moon, Mars, Earth orbital and free space (actually solar orbit), and other worlds a different model must prevail, at least for the first few decades of development. Instead of specialized factories mass producing many copies of a limited number of products every day, the job will require factories capable of producing a limited number of copies of a wide variety of different products and a large selection of them.

Local resources must be used for raw materials. This is called In Situ Resource Utilization or ISRU. On the Moon, regolith will be the main resource. There are basically two kinds of regolith. Basaltic mare regolith is composed mostly of iron and magnesium silicates. Highland regolith is made of aluminum and calcium silicates. On Mars, regolith will be a source of metals including iron and copper, chlorine and water. The atmosphere of Mars will provide carbon and nitrogen. Various devices will be imported from Earth to tap these resources. There will be machines resembling huge mass spectrometers or calutrons to separate metals and oxygen from regolith. There will be ovens to simply roast solar wind implanted volatiles from lunar regolith and water from martian regolith. These ovens might use direct solar heat or waste heat from nuclear reactors. There will be retorts for extracting metals by way of processes well known on Earth, like silicothermic reduction of magnesia to produce magnesium.

There are other resources like solar energy, free vacuum or low pressure, low gravity, super cold with shielded space radiators and perhaps unpredictable resources yet to be discovered. Free vacuum will make large mass spectrometer like devices possible. Super cold temperatures will make it possible to easily liquefy hydrogen, oxygen and other gases. Solar energy never obscured by clouds is abundant on the Moon. In polar regions it is available up to 90% of the time and at lower latitudes 50% of the time. On Mars, solar energy is less intense and sometimes there are dust storms, but the day-night cycle is just about 24.5 hours long and that will be ideal for agriculture in plastic domes. In GEO, solar energy is available 100% of the time. In LEO a 45-50 minute light to 45-50 minute dark cycle must be dealt with. High orbit is a much better place for manufacturing. In free space or solar orbit solar energy is available 100% of the time.

Once we have metals, basalt, glass, cement, various gases and organic chemicals, what do we do with them? How do we make hundreds if not thousands of different parts and products in the limited number of copies we need? This seems to be a job for 3D printing. Metals that are either powdered or drawn into wires can be fed into Electron Beam Additive Manufacturing devices that work with the free vacuum or Selective Laser Sintering devices. This might be a great way to make steel dies for forging presses that can stamp out many different kinds of parts by switching dies. It could be the right way to make dies for injection molding of molten basalt or glass on the Moon or basalt, glass and plastic on Mars. Injection molding is a good way to mass produce things. The 3D printers tend to be slow but they could be used to make the steel dies and other parts for the injection molding machines. Human workers and robots could assemble the parts. This way of doing things could save lots of imported cargo mass and cut costs.

Basalt is an excellent resource. The seas of the Moon are filled with it and there is basalt on Mars too. It can be melted at about 1250 C. and poured or injected into iron molds. If we want to make large blocks or slabs of basalt it should be possible to simply dig sand molds in the ground. If we want to make bricks, tiles, small and medium sized parts, dinnerware, and more we will need to melt and cast it in iron molds that could be made by 3D printing with iron mined on site. Basalt can also be drawn into fibers to make fire-proof fabric for protective outerwear, curtains, rope, cables, furniture, filters, drop-cloths, etc. Basalt fibers can be three times stronger than steel in tension. Cables made of this stuff could be used to pre-stress structures that must support a lot of weight or pressure. Basalt rods can make rebar in concrete. Instead of shipping tons of iron molds into space we will use local resources and 3D printing to make them.

Basalt could be a very important base material. It is harder than steel and abrasion resistant. It is strong in compression but not so strong in tension and it is rather brittle. Uses for basalt include: [11]

## Cast basalt

machine base supports (lathes, milling machines), furnace lining for resources extraction operations, large tool beds, crusher jaws, pipes and conduits, conveyor material (pneumatic, hydraulic, sliding), linings for ball, tube or pug mills, flue ducts, ventilators, cyclers, drains, mixers, tanks, electrolyzers, and mineral dressing equipment, tiles and bricks, sidings, expendable ablative hull material (possibly composited with spun basalt), track rails, "railroad" ties, pylons, heavy duty containers for "agricultural" use, radar dish or mirror frames, thermal rods or heat pipes housings, supports and backing for solar collectors

## Sintered basalt

nozzles, tubing, wire-drawing dies, ball bearings, wheels, low torque fasteners, studs, furniture and utensils, low load axles, scientific equipment, frames and yokes, light tools, light duty containers and flasks for laboratory use, pump housings, filters/partial plugs

## Spun basalt (fibers)

cloth and bedding, resilient shock absorbing pads, acoustic insulation, thermal insulation, insulator for prevention of cold welding of metals, filler in sintered "soil" cement, fine springs, packing material, strainers or filters for industrial or agricultural use, electrical insulation, ropes for cables (with coatings)

More everyday items that could be made of cast, sintered or fiber basalt include plates, dishes, mugs, tea cups, bowls, tea and coffee pots, serving trays, pitchers, decanters, counter tops, kitchen sinks, table tops, table legs, stools, chairs, bars, shelves, bottles, jugs, hand basins, toilets, bath tubs, shower stalls, bidets, planting containers, flower pots, vases, lamps, water pipes and sewer pipes, ash trays, paper weights, candle sticks, aquaculture tanks, floor, ceiling and wall tiles, bricks, blocks, towel racks, clothes racks, shower curtain racks, shower curtain rings, shower curtains from basalt fiber, drapes, cushions of woven fiber stuffed with fiber, mattresses, rugs, statuary, doors, handles and knobs for doors and drawers, picture frames and certainly other things. Small items could be cast or sintered in 3D printed iron molds while large items like toilets and aquaculture tanks could be cast in expendable sand molds bound with sodium silicate.

Basalt might also be used to make habitat. It melts at around 1150 C. to 1350 C. and could be extruded from a machine similar to a contour crafting gantry used to print buildings on Earth from cement.

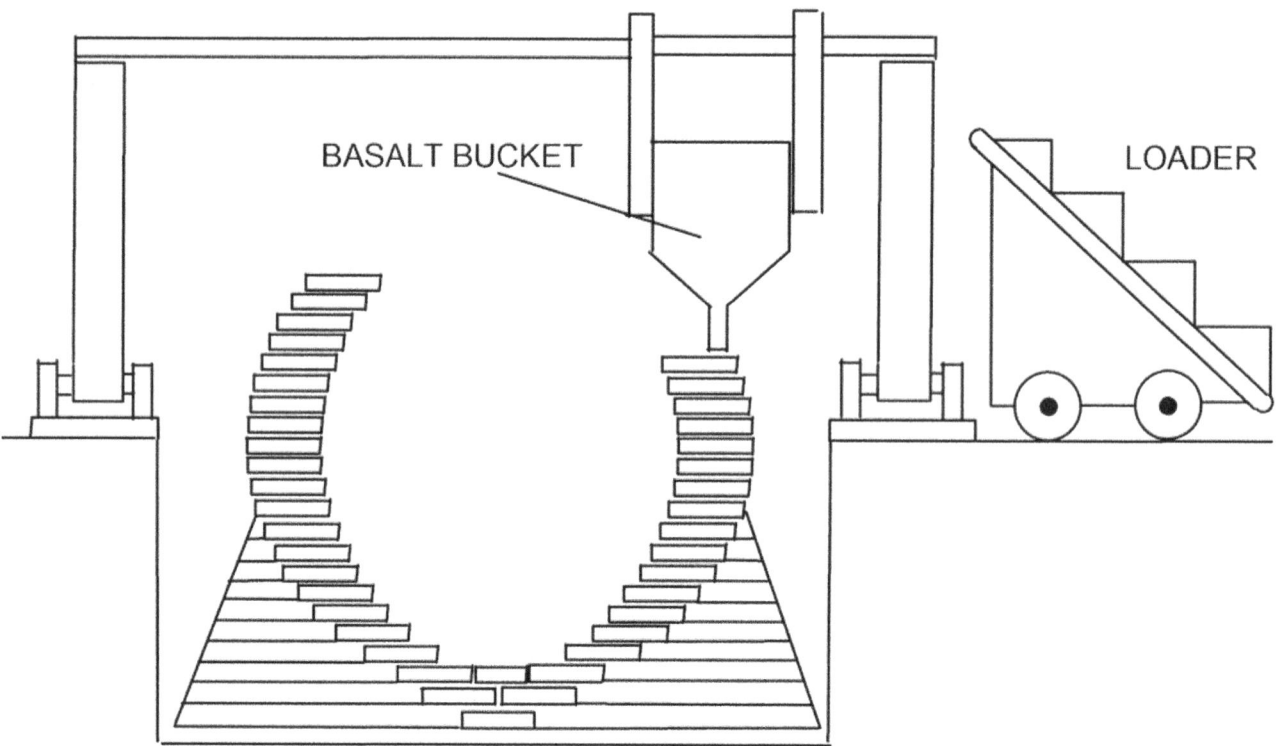

Fig. 5 Cross sectional diagram of machine printing cylindrical habitat with extruded molten basalt.

This material will have no trouble standing up to the heat of the lunar day. However, thermal cycling between night and day could lead to cracking. Layers of loose regolith several meters thick would cover the structures to provide cosmic ray shielding and more than enough thermal insulation that would prevent thermal cycling. Basalt doesn't have a lot of tensile strength compared to metals so module walls might be as much as a foot thick as opposed to mere fractions of an inch that would be needed with metals. Basaltic mare regolith would be easy to excavate, screen, and load into contour crafting gantries. No metal extraction, metal working and welding would be involved. This could be a much cheaper way to make habitat modules and it is easily automated. Energy from solar panels will be needed to melt the material but that shouldn't be a showstopper.

Electric motors have been made by 3D printing. Researchers at Chemnitz University of Technology have printed an electric motor from copper, iron and ceramic.[12] The motor can work at high temperatures of about 300 C. These motors will be needed in large numbers at all space facilities.

What about large machines? Rolling mills for sheet metal, plates, rails, beams, etc. can weigh ten, twenty or more tons. Several of them in series will be needed to reduce slabs of metal to thin plates or sheets. They could be built on-site. The large, heavy rolls and frame parts could be made of steel produced from local iron and carbon. These parts could be cast in silica sand molds bound with sodium silicate on the Moon and possibly resins on Mars. The parts could be finished with CNC machines in pressurized inflatable modules where lubricants won't simply evaporate. Smaller parts, like bearings and bolts, could be made with 3D printers or by humans in machine shops equipped with lathes, drill presses, mills, bolt rolling machines, etc. Extruder machines could be made by sand casting and CNC machining. A large lathe might be used for horizontal boring and spinning large domes from metal plates.

Humans with machine tools could make almost anything metal by hand. They could even make more machine tools and replicate the shop. The only problem with humans is that they require air, food, water, mild temperatures, habitat, spacesuits, creature comforts and paychecks with insurance benefits. Will the mass and cost of all this be worth it or should humans be kept out and robots used entirely? Somehow, I doubt that robots could handle everything but advances in AI could surprise us all.  At this time, it seems like humans and robots working together will be the way to do things.

On demand manufacturing instead of mass production will probably be used.  Mass production creates huge inventories that must be warehoused until sold.  The only customers we will have in space will be ourselves.  It would make sense to only make what we need when we need it in required numbers and have a few spares lying around.  Eventually large numbers of tourists will visit the Moon and it might make sense to have retail stores with well stocked shelves.  The same could be true at Mars settlements and in large space habitat.  By using ISRU and making things to make the things needed for making things it will be possible to expand industry.  This is called bootstrapping.  Given enough time, growth, tourists and settlers, there could be a demand for large numbers of ground and sub-orbital space vehicles and railroad cars.  There could also be a large demand for mundane things like toothbrushes, razors, coffee mugs, clothing, etc.  A shift to mass production might happen when it becomes possible.  The original bootstrapping mining bases on the Moon, Mars and in GEO when lunar materials are made available could grow into large scale industry consisting of hundreds of factories or more.  There could also be lots of small shops run by settlers to keep the shelves in small stores and boutiques stocked up. Some people might make candles, polished Moon rocks, soap or metal knives as a sideline.

Manufacturing in space proper, like GEO, will depend on resources from the Moon and eventually asteroids.  The machines needed to make the machines to make the things needed for humans and more machines will be rocketed up at first from Earth.  In GEO, lunar regolith could be used to bootstrap industry for building SSPSs and rocket propellants.  Steel and other materials could be hauled down to LEO where large habitats like Kalpana Two are built. Beyond Earth orbit asteroids will be the main source of raw materials.  Large space settlements could be built in the proximity of NEOs and Main Belt Asteroids.  They will probably start out with a limited amount of equipment and use the same bootstrapping techniques used on the Moon and elsewhere to expand until their industry could use asteroids to build settlements of great size.

Asteroids will also be of great value in Earth orbit. Rocket propellant will be in demand in GEO and LEO. Lunar regolith can provide plenty of liquid oxygen which comprises the bulk of most chemical propellant combinations. Plain regolith can be used for mass driver propulsion or ablation propulsion which uses lasers or electron beams to vaporize material and create rocket exhaust. Lunar sodium or magnesium might work with ion drives. Hydrogen could be "piggy-backed" with cargo and passenger flights to LEO and it could be extracted from lunar polar ice. The Moon's ice is a finite resource that could last for centuries if it is used for water that is recycled endlessly. It would be a crime against Nature to use up all the ice for rockets that spew it into the vacuum of space and lose it forever. Some ice could be sacrificed in the early days of lunar industrialization until a foothold was established in space, but this cannot and should not go on forever. There are thousands of asteroids that contain water and hydrogen bearing organic compounds. Near Earth asteroids should be tapped for their resources of water and organics long before much lunar ice is used.

Manufacturing and construction in LEO presents a challenge. Solar energy is available for about 45-50 minutes followed by darkness for 45-50 minutes as the spacecraft orbits the Earth. Space stations in LEO where large habitat, propellant depots and spaceships are assembled will need extra solar panels and batteries to provide power during darkness. Another resource besides lunar steel, cement and glass that's of interest in LEO is external tanks. If a rocket based on a booster and an ET with a reusable aerospike engine module is used, then every cargo or passenger flight would provide the added benefit of about 30 metric tons of aluminum, copper, lithium, titanium and polyurethane for use in LEO. This is in addition to a payload mass of 100 or more metric tons. The tanks themselves could be modified for use in orbital propellant depots. They could be cut up and melted down. The metals could be separated electrically or chemically and used for making some of the many parts of space stations and ships in LEO. Much of the same kinds of machines used for bootstrapping on the Moon and in GEO could be used. To be really effective, the tanks would have to be mass produced on largely automated assembly lines to bring their cost down from about $60 million to just a few million dollars or less.

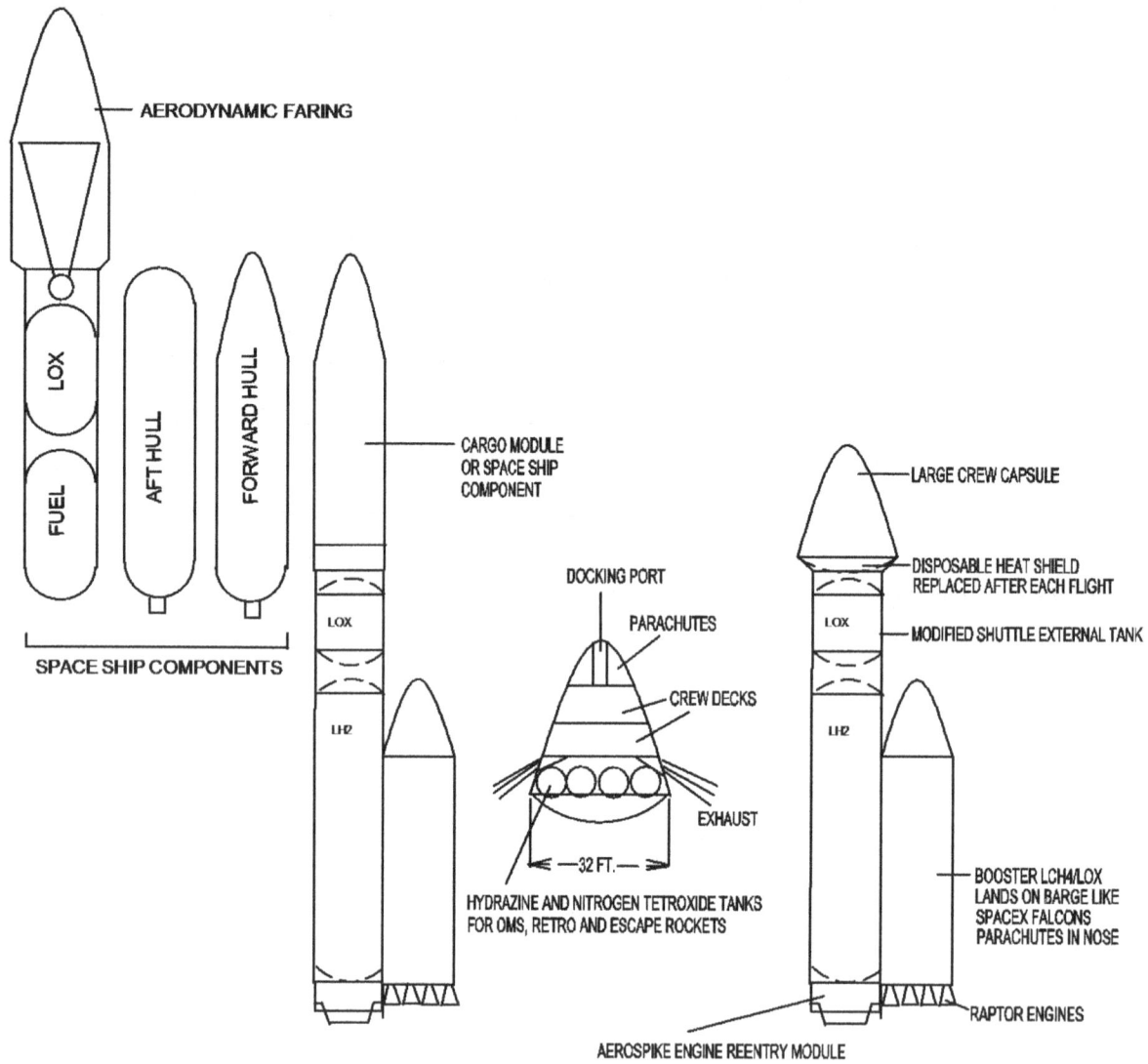

Fig. 6 Space Shuttle External Tank based rocket system.

The rocket pictured above could transport passengers or cargo to LEO. The external tanks and cargo modules could be used in space for construction and manufacturing. Spaceship components, each as complicated and pricy as a jet airliner, could be orbited and assembled at space stations. The side mounted booster would land on a barge like Falcon rockets do today. The aerospike main engine module returns to Earth for reuse. Since the tanks, cargo modules and aerodynamic farings are utilized in space, the system is essentially 100% reusable. The cost of manufacturing tanks will determine the economics of this system.

# Interplanetary Trade

Every nation on Earth trades with other nations. It seems that no nation can produce everything its people need and want, be it raw materials or finished products, or both. Some nations are rich with natural resources and other nations have lots of cheap labor. A similar situation is likely to exist in space. The Moon has lots of raw materials but it lacks many needed elements. Substitutes will be found or these elements will be imported from Earth, Mars or asteroids. Asteroids might supply Earth, the Moon and Mars with platinum group metals. The Moon and orbital space might import hydrocarbons from asteroids. Orbital space industries will sell energy, real estate, vacations and spaceship transportation. The Moon will sell vacations and travel also. Mars presents a problem. What can Mars sell to the Moon that can't be obtained locally or from asteroids? What can the Moon sell to Mars?

The first order of business will be the establishment of a base on the Moon that bootstraps itself up to a point at which millions of tons of raw material are launched into space every year. This will cost ten or twenty billion dollars based on my estimates.[13] The next order of business is the establishment of GEO work stations where materials are processed and SSPSs are built. These stations will be made by assembling parts rocketed up from Earth and more will be bootstrapped once lunar materials are available. This will cost about as much as the Moon mining base.

The SSPS builders will pay for lunar regolith shipments. A 5 GWe powersat might have a mass of 50,000 tonnes. Regolith is 40% oxygen and 60% silicon and metals. Builders in GEO won't need much oxygen except for atmospheres and high pressure cold gas thrusters for moving Manned Maneuvering Units (MMUs) and Orbital Maneuvering Vehicles (OMVs) around. Aluminum, magnesium and titanium will be the main construction materials for powersats. Small amounts of silicon, manganese and chromium will be needed for alloys. Some glass might be useful but not a lot of iron and calcium will be needed for SSPSs. Instead of silicon for solar panels, multi-junction solar panels will be sent up from Earth and the powersat will use reflectors to concentrate the Sun's rays on them. It is safe to say 200,000 to 250,000 tons of regolith will be needed for one

50,000 ton satellite. An equal amount of regolith might be needed for reaction mass for mass catchers, therefore the Moon base has to excavate 400,000 to 500,000 tons of regolith. That would be about a third of a square kilometer dug to depth of one meter. Mass drivers have been envisioned that can launch 600,000 to 6,000,000 tons of material every year.[14] If the Moon miners charge two dollars per kilogram of delivered regolith they will earn $400 million to $500 million before costs per SSPS and the GEO builders will pay that much.

A five cents a kilowatt hour for wholesale electricity a 5 GWe powersat could rake in $2.19 billion a year. It looks like the cost of building these can be recouped in a short amount of time. Several construction space stations will be built and perhaps ten SSPSs could be erected every year. That would require four million to five million tons of regolith per year or about one square kilometer dug 3 meters deep. The Moon base could earn $4 billion to $5 billion per year. In a few years it could break even.

In 100 years there could be a thousand powersats delivering a total of 5 TW of electricity or 43.8 trillion kilowatt hours per year worth $2.19 trillion. This could supply a very significant portion of future energy demand. What about all that excess regolith and oxygen from metal smelting? There would be 1.5 to 2 million tons of this to deal with every year if ten powersats are built every year. The iron, silicon dioxide (glass) and calcium oxide (lime) could be sold to LEO settlement builders and rocket propellant companies. At four dollars per kilogram to cover the cost of buying the stuff from the Moon in the first place and the cost of extracting the desired metals, this could be worth $6 billion to $8 billion per year.

A space settlement in equatorial low Earth orbit (ELEO) like Kalpana Two would have a mass of 16,800 tons.[15] If built mostly out of steel and glass instead of aluminum it might weigh more. Say it weighed 25,000 tons. At $4/kg. that would cost $100 million. Reasonably priced real estate, for millionaires and billionaires at least, could be created in Earth orbit at this price. Hundreds of thousands of tons of iron and glass could be extracted from 1.5 to 2 million tons of excess regolith every year. There would be more than enough to build space settlements and have a lot left over for rocket propellant.

If a lunar transit company had ships that used a thousand tons of metals and oxygen propellants per round trip to the Moon, that would cost $4 million at four dollars per kilogram. If 200 passengers each paid $100,000 for a round trip to the Moon, that would be $20 million or $16 million over fuel cost. After other expenses, there is certainly to be some decent profit. If the ship can make one round trip per week, with two weeks for refurbishment every year, it can make 50 flights a year and earn $800 million over propellant costs. The ship might cost a billion dollars or more, but it could pay itself off in a few years and make clear profit after that.

It looks like trade between the Moon, high Earth orbit and low Earth orbit would be worthwhile. Space Solar Power Satellites look like big earners. Moon mining might not be as lucrative as selling electricity, but it would launch the space tourism industry which could become a big money maker. Lunar hotels and resorts must be built on the Moon to accommodate visitors. The mining base could replicate itself several times over and increase production until big construction projects were conducted on the Moon. Factories to build small vehicles, lunar suborbital and orbital spacecraft and railroad cars would emerge.

Robotic orbital spacecraft could clean up space junk including old satellites that contain precious metals. The real gold mine would be near Earth asteroids. With materials from the Moon, propellants and orbital construction stations it will be possible to build robotic spaceships that travel to asteroids, mine them and bring payloads of platinum group metals, other precious metals, water and hydrocarbons for sale to space industries and the Moon. Some of those precious metals would also be sold to Earth.

Asteroid mining ships with AI computer brains could use nuclear power to energize lasers or electron beams to vaporize regolith for ablation propulsion. Unlike mass driver propulsion, this won't create navigational hazards from ejected bags or blocks of regolith. A mass driver engine would be about a thousand feet long and require magazines full of bags of regolith or billets of sintered regolith. Ablation propulsion units should be more compact. Upon reaching an asteroid the ship would deploy mining machines that dig up the dust and rock, crush it up and roast it to drive out water and tarry hydrocarbons. The spent dust and rock would be sintered into a large block and used as reaction mass for ablation propulsion.

The ship would leave mining and processing machinery at the asteroid to do more work. It would just haul the payload of tar and water back to Earth orbit or the Moon. This material would be used for life support and chemicals to make drugs, plastics, paint, etc.

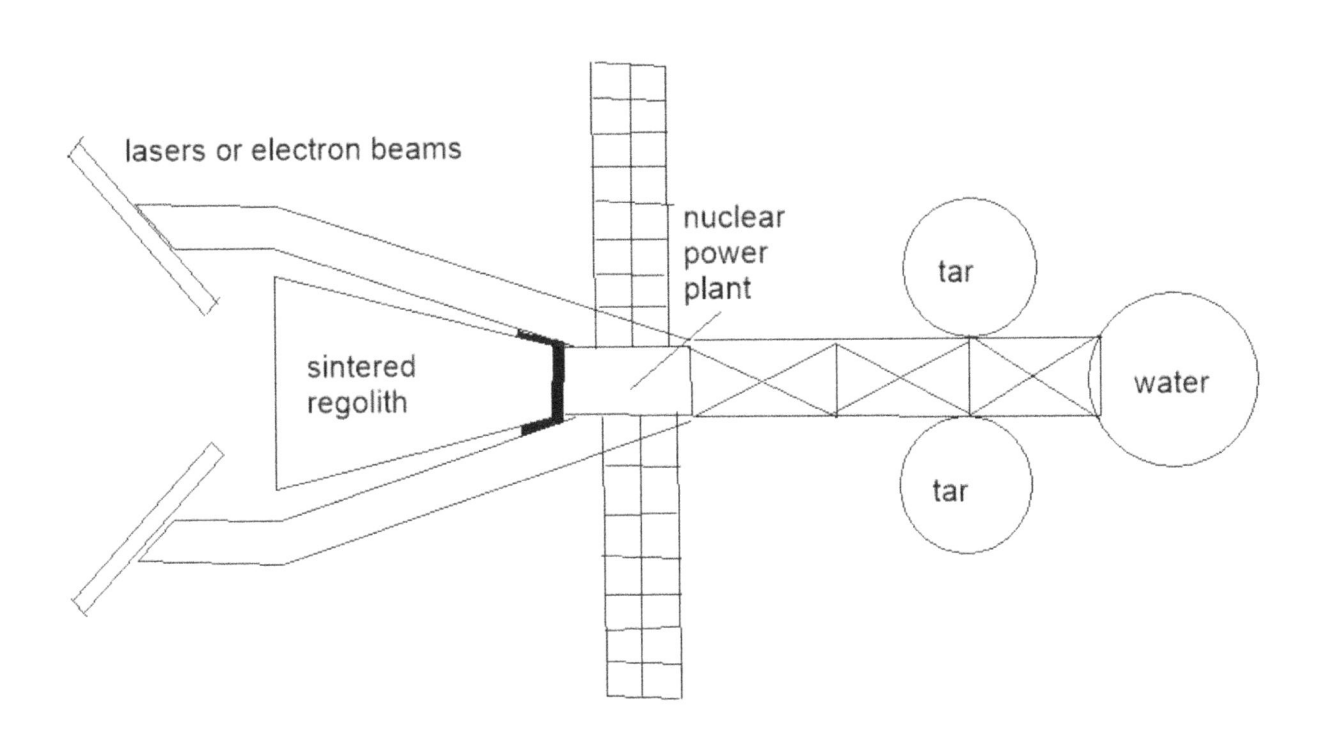

lasers or electron beams

nuclear power plant

tar

sintered regolith

water

tar

robotic freighter/tanker with ablation propulsion

Fig. 7 Asteroid mining ship

Mining metallic asteroids consisting mostly of iron and nickel will require high powered lasers to cut pieces of the solid metal away from the asteroid, sort of like slicing up a cake. Magnets would pull the pieces away and send them to processors that melt and powder the metal, use magnetic and other separation techniques to extract the platinum and other precious metals, and leave the iron behind. Nickel might have more value than iron. Besides platinum and gold, there might be some germanium and gallium for making multi-junction solar cells.

The asteroid mining spacecraft will be robotic for a simple reason. Although it takes less delta velocity to reach many near Earth asteroids than it does to travel from Earth orbit to the surface of the Moon, the launch windows to and from these asteroids can be years apart. It would be very difficult to sustain humans in space that long.

With lunar and asteroid resources space industry could grow by leaps and bounds. There would be plenty of hydrogen for nuclear electric propulsion systems and metals for building large space liners for travel to Mars and other destinations in the solar system. Many settlers might have already gone to Mars before asteroid mining is big. They might go as soon as lunar materials are available for propellant in spaceships built on the ground and rocketed up to orbit. The Martian settlers will have to produce everything they need from resources on Mars because of the cost of shipping things to Mars. There isn't a lot that Mars can offer to Earth orbit and the Moon. There could be vacations for the wealthiest travelers that earn enough for the Martians to buy some items from Earth. Mars has copper and chlorine that have many industrial uses. It actually takes less energy to travel from Mars to the Moon than from Earth. Mass drivers atop an equatorial volcano like Pavonis Mons could launch payloads into Mars orbit that are then hauled with magnetic or light sail robotic spacecraft to the Moon. It's also possible that Mars contains ore deposits of sedimentary origin that formed billions of year ago when Mars still had liquid water. The Moon's KREEP terrains have low concentrations of uranium and thorium. If richer ores of these nuclear fuels are found on Mars it might actually be cheaper to import uranium from Mars than produce it on the Moon.

The settlement of Mars is unlikely for financial reasons, unlike the industrialization and settlement of the Moon and Earth orbit. The only profits Martians can expect are from trading with other Martians. In thousands of years, Mars might be terraformed and the descendants of the original settlers might claim territory and sell it to immigrants who want to live there. The lure of Mars is great. Terrestrial ex-patriots might go there just because they want to and they may be willing to spend all their money to go there and stay there. There will certainly be plenty of space businesses willing to sell them passage by spaceship, survival equipment, tools, spacesuits, solar panels, provisions, electronics, vehicles and anything else they will need to survive and grow on Mars.

# Breaking it Down: Process Summary

1) Up to 1000 tons of equipment sent to the Moon. This will include multi-junction solar panels, inflatable habitat, excavators, super-sonic dust roasters and all isotope separators, several kinds of 3D printers, mass driver lunar launcher sections, electric and solar ladle furnaces for melting and pouring materials, CNC machines, etc.

2) Moon base "bootstraps" and grows using local resources and starts launching loads of regolith to mass catchers stationed at L2. Mass catchers haul large loads of regolith to GEO.

3A) Rotating GEO stations are assembled with parts from Earth. Inflatable habitat is included for human crews. These stations are fitted with super-sonic dust roasters and all isotope separators to produce metals. Lunar sourced iron is converted to steel with carbon from Earth. Steel is sand cast and machined with CNC machines. Parts are made with 3D printers. Heavy rolling mills are assembled from steel parts. Aluminum and magnesium alloy sheet metal is produced in large quantities. Titanium parts made by 3D printers for high stress applications.

3B) By-product oxygen resulting from metal extraction is cooled with space radiators, compressed, liquified and stored in tanks derived from upper stages or external tanks, or made on site from spun aluminum (requires large lathe, etc.). Gaseous oxygen in high pressure tanks used for cold gas thruster propulsion of various manned and robotic orbital vehicles.

3C) By-product silicon, iron and calcium silicate slag in excess of that needed for power satellites construction is stored on non-rotating weightless platforms.

4) Beam builders are sent to GEO stations. Sheet metal is used to make large lightweight beams that are assembled into SSPS frames. Sheet or foil reflectors are installed by robots; some autonomous and some teleoperated. Beam builders might also be made at least partially on-site with lunar sourced metals, 3D printers, CNC machining, robot and human workers.

5A) Multi-junction solar panels sent to GEO and combined with frame and reflector assemblies for concentrating solar energy systems. Parts of

Amplitrons are sent up. Microwave transmitting system and antenna assembled from wave guides and dish made from sheet metal. Flywheels with electric motors installed to control satellite's attitude. Propulsion system might also be needed to keep satellite on station due to drifting resulting from gravitational perturbations caused by the Moon and other solar system bodies.

5B) More GEO stations are built using lunar resources. The stations consist largely of aluminum beams made by rolling or extrusion. More power satellites are built.

6) ELEO space stations and orbital propellant manufacturing and storage stations are established. Solar electric tugs that use a combination of electrodynamic tethers and ion drives haul loads of silicon, iron, slag and oxygen from GEO. LOX is stored in depots. Silicon is combined with hydrogen "piggybacked" on rockets from Earth to make silane which is compressed, cooled in space radiators and liquefied.

7) Spaceship sections rocketed to ELEO and assembled. Loaded with LOX, silane and metal powders for propulsion to lunar orbit.

8) ELEO space stations used as bases for construction of large Kalpana settlements. This requires steel made from lunar iron. Glass and cement are also produced. Silicon can be re-oxidized to make $SiO_2$. The separators in GEO produce CaO and MgO. This mixture can be reduced with silicon to get magnesium metal for power satellites and calcium silicate slag that can be used for cement. Aluminum from external tanks can also be used for space construction.

9) Back on the Moon, expansion has continued. Polar ices are mined. Rails are made by rolling and railways built. Remote solar power plants are built and cables are strung on basalt poles to provide mass driver base with power 24/7. Light elements from ice are launched into space. Habitations for large numbers of tourists are built.

10) Development in space continues using lunar resources, external tanks and products from Earth. Asteroid mining and Mars settlement begin. Asteroid resources exploited. Exploration of Mercury and the outer solar system begins followed by settlement.

# Laser Telecommunication

The sailors of old were out of contact with home for months even years at a time.  In today's world and tomorrow's solar system, space travelers and settlers will want to communicate with other worlds, space settlements and ships out in space.  Interplanetary distances are so great that it will take minutes to hours for electromagnetic waves to cross those distances.  People will send messages in the form of text, fax, voice and video to each other and just wait for a reply, much the way we do "playing tag" today via email.  Computer systems will store and forward messages.

Relay stations in space will be needed to direct and amplify signals to planets, colonies and ships in space. Relays will be needed when planets or ships are on the other side of the Sun.  Perhaps they will be located at Sun/Planet Lagrange points.

Lasers will probably carry data at very high rates.  Bandwidth or data rate in bits per second depends on signal to noise ratios. Larger signal to noise ratios allow faster data rates.  Since lasers don't spread out (beam divergence) as much as radio wave beams do over vast distances it should be possible receive stronger signals and achieve higher data rates.  This is really important when you are sending videos or vast amounts of data including financial data over tens and hundreds of millions of miles.

Laser beams will be effected by atmospheric absorption and distortion on Earth, Mars and Titan.  Clouds could block the beams entirely.  On worlds with atmospheres it will be necessary to have laser stations in orbit with radio links to ground networks.

Aiming lasers will be difficult. Tiny deviations at the transmitter could cause the beam to miss the receiver by many miles.  Ultra-high precision equipment will be needed.  Since the stations, planets and ships will all be in motion aiming becomes more complex.  Computers will use velocity and location data from inertial navigation systems and GPS along with trajectory data for the targets to control aiming.  Relay stations will need precise attitude and station-keeping control with reaction wheels and thrusters.

Beam divergence is directly proportional to wavelength and inversely proportional to aperture.[16] So the smaller the wavelength the tighter the beam and the larger the aperture the tighter the beam. Microwaves have wavelengths of 1 mm to 30 cm. An infrared laser might have a wavelength of 800 nm and a UV laser 300 nm. Although a microwave beam might be transmitted by a huge dish antenna (large aperture) its wavelength is so much larger than light from a laser that it will diverge much more and the signal will dissipate and weaken. The data rate is also dependent on frequency with higher frequency lasers capable of transmitting more data faster than radio can.

Relays will also be needed for distant ships. The relay stations could host large lasers and banks of telescopes to receive signals while ships might not be able to carry as much mass in the form of lasers and scopes. So it seems relays that can boost signals will be needed for consistent contact with ships.

You can use matrices to determine the shortest distance between a set of points. I think that's how GPS receivers/navigation systems determine the shortest route. It should be possible for computers to figure out what the best route from planet to relays to ship the beam should travel on. That might not always be the shortest route but it may be the one that can maintain signal strength and data rate at an acceptable level. The speed of light is one limitation, but weak signals and painfully slow data rates might be worse.

The relay stations could be staffed by AI robots that have no need for food, oxygen, sleeping quarters, recreation, hundreds of tons of cosmic ray shielding etc. This will mean the relays can be much less massive. Humans would not be happy at a deep space station anyway. Even at Lagrange points stations may drift. Station keeping could be done with ion drives or VASIMR which doesn't have electrodes to burn out and can use almost anything for reaction mass--hydrogen, inert gases, metal vapors, maybe even Scotch whiskey. Stations in the outer system might need nuclear power or some really huge reflectors to concentrate sunlight onto PVs. The AI might be smart enough to do repairs, but lots of redundancy may be needed for reliability during the lifetime of the relay station.

The outer solar system is enormous compared to the inner solar system. The space between Mars and Jupiter, the asteroid belt, is over three times wider than the distance of the Earth from the Sun. I suspect there would have to be relays in solar orbit and not just at Lagrange points. The inner system would probably be colonized first and that would allow time to develop better telecommunication technology like more powerful lasers and more sensitive optical receivers when humans settle the outer system.

Imagine the phone bills and ISP fees! I guess there would be lots of telecommunication companies involved with each one owning and operating different parts of the network and connecting. This is big business today. As humanity spreads out into space an even bigger market will evolve.

Routers, servers and a million algorithms. Building a seamless interplanetary telecomm system will keep armies of engineers and computer scientists busy for a long time.

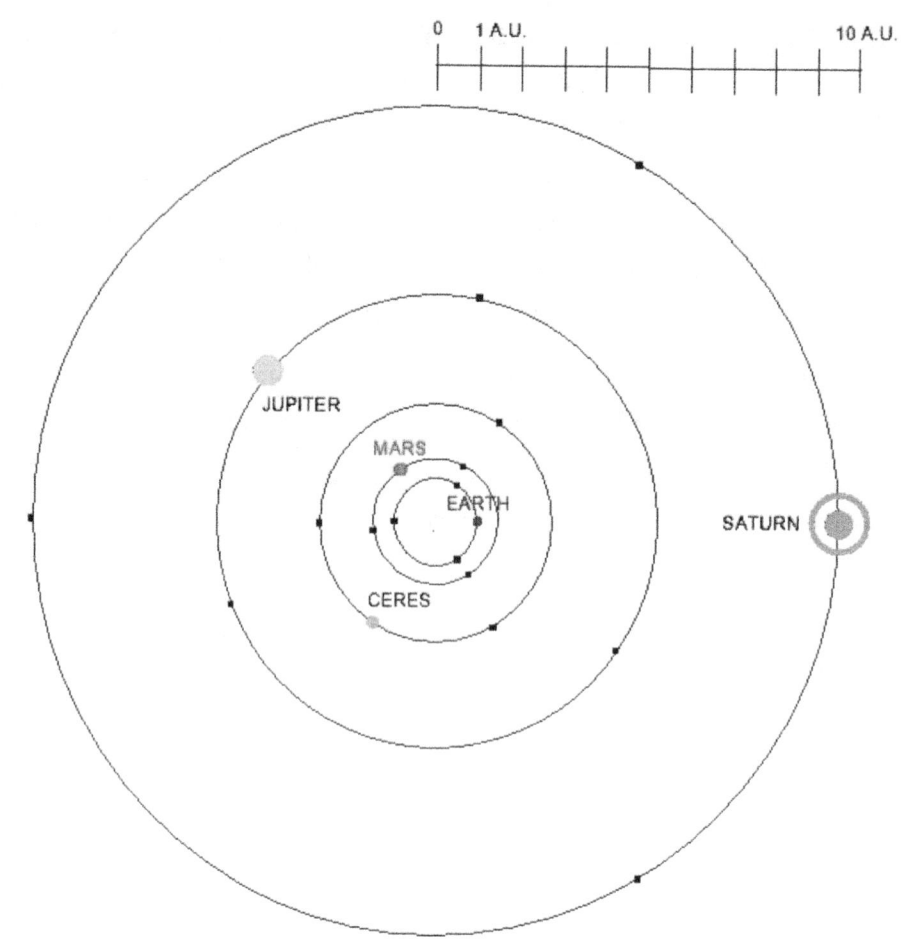

Fig. 8  Planets and Lagrange points where relays can be located.

# Pulsar Navigation

Over the years I have heard people suggest the use of pulsars as navigation beacons for ships in space and even ground vehicles on Venus. Well, NASA has already experimented with this. Experiments on the ISS with x-ray telescopes successfully located the space station's position.

Experiments on the ISS with the NICER (Neutron-star Interior Composition Explorer) x-ray telescope determined the station's position to within 3 miles. This experiment was called Station Explorer for X-ray Timing and Navigation Technology (SEXTANT). The NASA scientists predicted that they will eventually narrow that margin of error down to a few hundred feet.[17]

But wait a minute...isn't an x-ray telescope for tracking pulsars a massive piece of equipment?? The mirror on the Chandra X-ray telescope launched in 1999 had a mass of 18.5 metric tons per square meter.[18] The instrument used during the SEXTANT experiment only had a mass of 5 kilograms.[19] There has been a lot of progress in the field of X-ray optics. Pulsars emit in the radio portion of the spectrum, but a radio telescope attached to a spaceship would be rather large. Very few pulsars emit in the optical region (only five optical pulsars are known) and an impractically large optical telescope would be needed.[20] Compact X-ray telescopes win.

Presently, space probes are tracked and navigated by the DSN (Deep Space Network) using large dish antennas and radio. This won't suffice in the future when hundreds even thousands of ships are out in the solar system. To make things worse, the accuracy of radio tracking and navigation decreases with distance. Communication relay stations will need to know their location accurately so that lasers can be aimed precisely at them and at distant ships, colonies or worlds. Using pulsars for a sort of "solar system GPS" seems to be the wave of the future. The future communication laser beams will probably contain data pertaining to the location, trajectory and velocity of the ships and stations as well as text messages, email, faxes, voice and video messages and even internet searches and streaming content. Even in space we will want to stay connected and we definitely don't want our spaceships to get lost!

# Lunar Sports

Spectator sports are some of humanity's favorite past times. They are a multibillion dollar a year industry and have provided a way out of the ghetto for innumerable pro athletes. Besides ticket sales, sports create a market for everything from hot dogs and beer to T-shirts, caps and TV commercials. Sports on the Moon could be really exciting. How high could a basketball player jump? What would low gravity football be like? Ice hockey probably wouldn't be much different in low lunar G but tennis could get interesting. How far could a baseball player hit the ball??

Wait a minute...don't we need a stadium? How are you going to build a stadium in the vacuum from tens of thousands of tons of concrete and steel? Seems impossible. However, Nature has provided for us in the way She provides in so many ways be it x-ray pulsars for navigation beacons or vast asteroidal riches in space. There are lava tubes on the Moon that are hundreds of meters in diameter. We know this because rille valleys are collapsed lava tubes and many of them are that big. Through our telescopes we see intact portions of lava tubes in the rille valleys of the Moon. Many of these are several kilometers long. Once the ends of the tubes are dammed with tons of rock and melted regolith, sealed and pressurized with oxygen which comprises 40% of the regolith, all we have to do is put seats on the inner slopes of the tubes and play ball! Tennis courts could be made with concrete or cast basalt tile surfaces. Nets could be made from cords made of basalt fibers. There isn't enough carbon and hydrogen to make Astroturf on the Moon so playing fields will have live grass. By day filtered sunlight will be piped in and at night microwave sulfur lamps could illuminate the fields. Live broadcasts from the Moon will rake in millions of advertising dollars. Lunar tourists will be thrilled! Even the half time shows will be more exciting when cheerleaders are tossed a hundred feet up in the air!

Building sports complexes within lava tubes and transporting teams to the Moon (one of the perks of being an athlete) will be expensive, but there should be plenty of profit in this. Terrestrial pros will want to play on the Moon and so will home grown athletes from various communities on the Moon. Tourists and locals will fill the stadium for reasonable ticket prices.

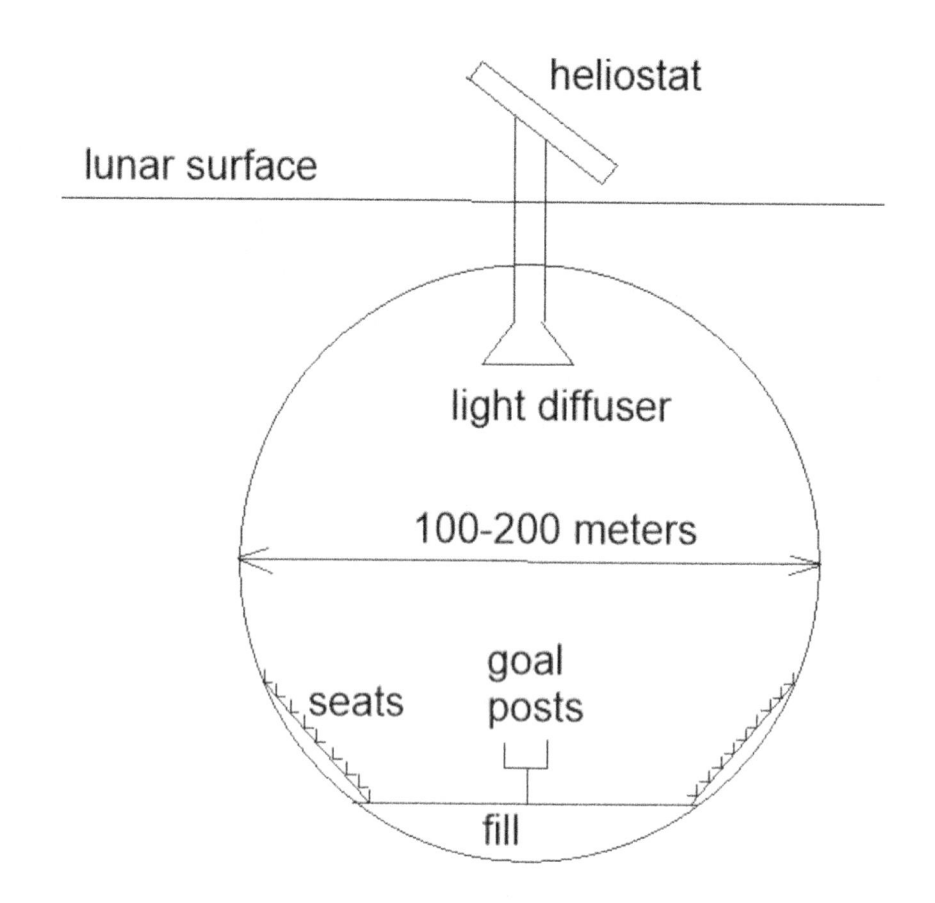

```
                        heliostat
                       /
   lunar surface      /
  ─────────────────┬─╱────────────
                   │
                   ▽
              light diffuser

            100-200 meters

                  goal
         seats   posts
                   ⊔
                  fill
```

## Lunar Lava Tube Stadium Cross Section

Circuses in low lunar gravity are another possibility. Large animals are not likely to be seen but aerialists and trapeze artists will do death defying things that they would never attempt on Earth. Dancers, showgirls and ballet stars could perform too. If a swimming pool can be constructed and enough water to fill it could be produced, divers could jump from spectacular heights. Trampolines could also be a lot of fun.

The open air environment of the lava tube(s) will make it possible to have real barbeques. Charcoal could be made from stems of crops that are ground up, compressed into briquettes and baked in electric ovens. Real beef for burgers might be rare and expensive, but hot dogs made from chicken and pork might be sold for a reasonable price. Tomatoes and spices grown on the Moon will provide barbeque sauce. Beer and other alcoholic beverages should be available along with soft drinks. The fun will literally be out of this world.

# Safety and Space

Frontier life is dangerous. That was true when early humans moved out of Africa and into temperate and even Arctic regions. It was true in the American West. Leaving the land and braving the seas in primitive boats was dangerous. The early days of air travel were dangerous. We cannot eliminate the dangers of space travel and settlement, but we can minimize those dangers. Space companies must take every foreseeable precaution or wrongful death suits will bankrupt them. Governments will establish regulations and organizations like a Federal Space Travel Administration that does for space what the FAA does for air travel.

Rockets explode too often. Major improvements in rocket reliability are needed if there is to be space travel for large numbers of people. A lot of research and development is called for. Perhaps performance can be sacrificed for reliability. This would reduce payload mass, but passengers are not that heavy. If the average traveler with luggage has a mass of 100 kilograms (220 pounds) then 100 people only weigh 10 metric tons. The SpaceX Starship is expected to have a 150 ton payload weight. If lower rocket engine thrust chamber pressures cause that to drop to 50 tons and this results in a significant increase in engine reliability then the orbiting of 100 people could still be done. The Starship and hopefully other rocket designs put the passengers on top of the first stage. The Shuttle mounted the orbiter with passengers on the side. With the vertically stacked design the chances of firing escape rockets and getting away from an exploding booster beneath are higher.

Unlike an airliner, spaceships cannot simply drop oxygen masks and descend to a lower altitude if cabin pressure is lost. It might be necessary for passengers to wear pressure suits. While this might not be as appealing as casual attire in "shirt sleeve" environments it might save lives. Space travel is not a walk on the beach. Passengers must be able to endure at least 3 Gees during ascent to orbit. They must know how to operate a space suit. They must understand emergency procedures. An instruction course that includes riding in a centrifuge and zero-G flights might be wise before leaving Earth. People with heart disease, severe diabetes, morbid obesity, high blood pressure and other major medical conditions would be smart not to try space travel. Many people might be motivated to lose weight, exercise and improve their health just so that they can take a trip into space. Before getting onboard a space vehicle a doctor's release may be required if only for reasons involving company liability.

Medical emergencies in LEO could necessitate a rapid return to Earth. This might not be a good idea at times since re-entry involves enduring negative G forces. Complete medical and dental facilities will be needed in Earth orbital stations and on the Moon. Nobody will want to cut a million dollar trip short just because of a bad tooth or a sprained ankle, much less something more serious. Surgery in weightlessness seems as if it would be impossible. How could bleeding be controlled? Ships to Mars will need centrifuges where surgery can be performed if needed. The Mars settlements will need complete hospitals too and well trained staff. Many people seem to hate doctors, dentists and lawyers until they need one. Medical personnel will be needed in space just as they are needed at sea.

Protection from and prevention of crime in space is also important. Some humans are dangerous and it's not always possible to weed them out. Cops will be needed on the High Frontier. Private security forces might be sufficient in LEO settlements where perpetrators can be apprehended and sent back to Earth for trial. On the Moon and Mars there will have to be police forces with guns, crime labs and judges. Lawyers might work with their clients remotely if the price of telecommunication is not exorbitant. Crimes like murder and sexual assault might be less frequent than drunken and disorderly conduct, peace disturbance, property destruction and petty theft. There could be mental health crises that require police or security guard intervention.

Mechanical and electrical systems in settlements and ships will need regular maintenance and sometimes repairs. A staff of technicians will have to be on hand. Since the crews on ships will be kept to a minimum, pilots and flight attendants will be trained in emergency repairs, equipment diagnostics, and emergency medical procedures. Hopefully there will be some doctors, nurses, engineers and technicians among the passengers who can help out if someone gets sick or the electrical system malfunctions. Most of the equipment will be monitored by sensors that can provide data to computers that locate malfunctions quickly. A smart ship will be a safer ship, and so will habitat in space and on the surfaces of other worlds. Crews will need rigorous training and certification by the Federal Space Travel Administration. Ships and habitat will also be routinely inspected and certified for spaceworthiness. Inspectors and cops will be paid for by license fees and taxes on the price of space travel.

# Thoughts About Space Settlement Construction

Large space settlements like O'Neill cylinders or Island 3, a cylinder 20 miles long and 4 miles (6400 m) wide, will be like steel balloons. Steel is the most likely construction material for large settlements because iron and nickel are abundant in asteroids while titanium and aluminum are not. The regolith of the Moon contains reasonable quantities of titanium and aluminum, but it is more likely that very large space settlements will be built of asteroid materials. Carbon is needed to make steel and a small amount of carbon can make a large quantity of steel. Sufficient carbon resources exist in C-type asteroids. Medium carbon steel with 0.35% to 1% carbon would be required to get metal with roughly 115,000 psi tensile strength. Large rods of asteroid iron would be packed in carbon and brought up to red heat for several days in solar furnaces. Carbon will dissolve into the iron and steel will form. Enormous furnaces will be needed to make steel in space. The steel will not rust on the outside in the vacuum of space. The inner surface of the hull would be coated with glass or aluminum to prevent rusting caused by moisture within.

Another challenge is heat treating the steel in space. To harden and strengthen the steel it will have to be heated then rapidly cooled or quenched. Enormous chambers in which large steel pieces are heated by concentrated solar energy, electric coils or hydrogen/oxygen flames then quenched with sprays of water will also be needed. Electric coils and water won't be compatible. It might be possible to heat steel in electric furnaces then shove it into a sealed chamber where it is doused with water.

Water exists in C-type asteroids in the range of 3% to 22% based on the study of meteorites. This water would have to be carefully recycled. The steam that forms from hydrogen/oxygen flames and quenching sprays must be recaptured and condensed.

Curved plates of steel for space settlement hulls might be 50 ft. by 50 ft. and 12 inches thick. They will be more like slabs than plates and giant rolling mills will be needed to shape them. Cold rolling will work harden the steel. This may or may not be desirable. Hot rolling will be easier on the rolling machines and it will not work harden or add stress to the steel parts, but it will require more energy. The machinery needed to build a giant space settlement will be almost as impressive as the settlement itself. Industrial capacity in space will start out small and smaller settlements will be built at first. In time the industry of space will grow and so will the settlements. Thousands of workers aided by robots will take years to build these structures.

A lighter structure might be made with a thinner hull reinforced by longerons and bands of high strength steel, especially in the windows. Glass fiber cables which can be stronger than steel might be used for the longerons and bands. They could be tightened up to pre-stress the hull so that the compression and tension forces cancel each other out to some extent. Bridges are built today with pre-stressed concrete using steel, glass or basalt fiber cables. Tensile stress on pre-stressed structures can be greatly reduced.

Electron beams can make welds 12 inches deep so two layers of steel could be combined to get a 24 inch thick hull. The layers could be charged with electricity to fuse them together. Asteroid iron contains 5% to 10% nickel and a trace of cobalt. Nickel increases the strength and hardness of steel without making it more brittle. It gives the steel fracture resistance. This is called toughness, and a steel hull that will resist fracturing (cracking) is of great importance. Chances are that a hull thicker than 24 inches will be needed. With a diameter of 6400 meters a hull one or two meters thick would still be comparatively thin, but it could be strong enough especially with pre-stressing. The stress due to the weight of the rotating hull, the soil layer within and internal air pressure must be examined.

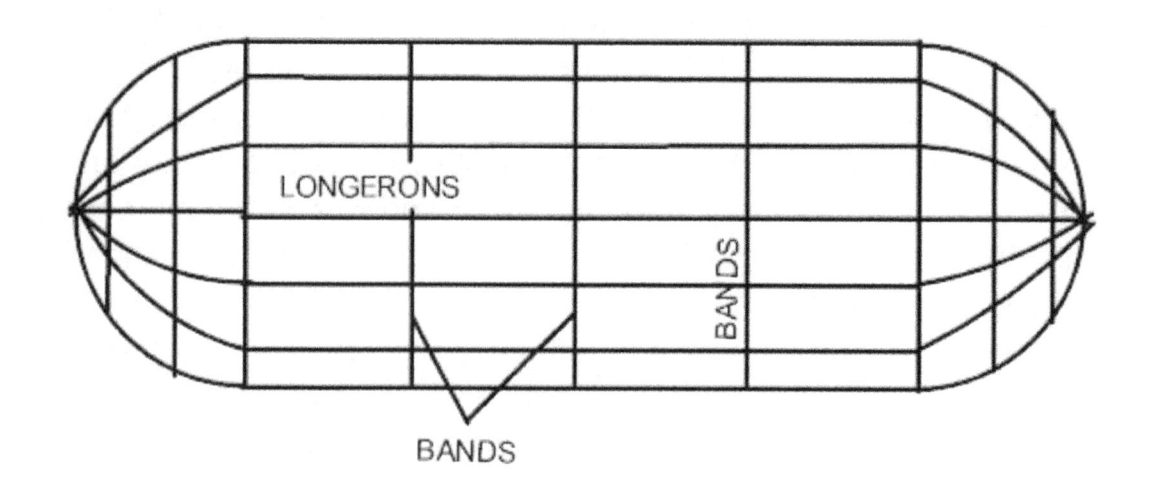

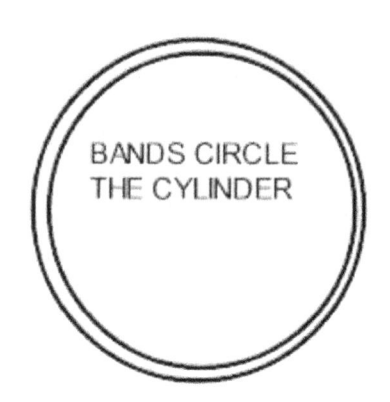

The circumferential or hoop stress on a rotating ring is given by: [21]

$$\sigma_z = \omega^2 \rho \, (r_1^2 + r_1 r_2 + r_2^2) / 3$$

$\sigma_z$ = stress (Pa, N/m²)

$\omega$ = *angular velocity* (rad/s)

$\rho$ = *density* (kg/m³)

where

$r_1$ = outer radius of ring (m)

$r_2$ = inner radius of ring (m)

Island 3 rotates once every 114 seconds, therefore 6.28 rad/114s = 0.055 rad/s = w

p = 7800 kg/m³   r1 = 3200m   r2 = 3199m if the hull is one meter thick
3198.8m if hull is 1.22m thick


$\sigma_z$ = 241,537,304 Pa = 35,032 psi 1m thick hull


$\sigma_z$ = 241,520,697 Pa = 35,030 psi 1.22m (48 in.) thick hull


psi = 0.00014504 Pa


As we can see, the hoop stress for a 1m or 1.22m hull is about the same. This stress could easily be endured by even mild steel, but we have to consider the stress due to the weight of soil and internal air pressure.


Hoop stress on a thin walled cylinder is given by:


σ = pr/t   p= internal pressure   t=thickness


A layer of soil weighing 11 metric tons per square meter of hull area will exert a pressure of 15.6 psi (107,556 Pa) and a 10 psi (68,946 Pa) atmosphere with 30% oxygen will exert a total pressure of 25.6 psi or 176,502 Pa. With a 1.22m (48 in.) thick hull the stress is given by:


σ = 176,502 * 3200/ 1.22 = 462,956,066 Pa or 67,147 psi


When added to the stress from the weight of the rotating steel hull, the total stress is:

*241,520,697 Pa + 462,956,066 Pa = 704,476,763 Pa or 102,177 psi*

With a 50% margin of safety an alloy steel with 153,266 psi would be needed. There are plain carbon, nickel and chromium steel alloys that strong. Pre-stressing with glass or basalt fiber cables will make it possible to use a steel alloy with less tensile strength; however, we might want very strong steel and pre-stressing for more safety, less anxiety and a longer life span for the settlement. Static pressure over a long enough time period could eventually lead to fatigue and failure of the hull, so it seems wise to use the best steel possible and pre-stressing of the structure.

| Steel | Type of Steel | Tensile Strength Kpsi | C | Mn | P | S | Si | Ni | Cr | Mo | V |
|---|---|---|---|---|---|---|---|---|---|---|---|
| 1045 | Plain Carbon | 80-182 | 0.43-0.50 | 0.60-0.90 | 0.04 max | 0.05 max | | | | | |
| 1095 | Plain Carbon | 90-213 | 0.90-1.03 | 0.30-0.50 | 0.04 max | 0.05 max | | | | | |
| 1330 | Manganese | 90-162 | 0.28-0.33 | 1.60-1.90 | 0.035 | 0.040 | 0.20-0.35 | | | | |
| 2517 | Nickel | 88-190 | 0.15-0.20 | 0.45-0.60 | 0.025 | 0.025 | 0.20-0.35 | 4.75-5.25 | | | |
| 3310 | Nickel-Chromium | 104-172 | 0.08-0.13 | 0.45-0.60 | 0.025 | 0.025 | 0.20-0.35 | 3.25-3.75 | 1.40-1.75 | | |
| 52100 | Chromium | 100-240 | 0.98-1.10 | 0.25-0.45 | 0.035 | 0.040 | 0.20-0.35 | | 1.30-1.60 | | |
| 9840 | Nickel-Chromium-Molybdenum | 120-280 | 0.38-0.43 | 0.70-0.90 | 0.040 | 0.040 | 0.20-0.35 | 0.85-1.15 | 0.70-0.90 | 0.20-0.30 | |

Longitudinal or axial stress is given by $\sigma = pr/2t$. This equation can also give us the hoop stress on a sphere. Some space settlements might be spherical. This stress is usually much less than hoop stress so we won't worry too much about it. Longitudinal stress will be caused by air pressure within and to a lesser extent the weight of the hull and soil layer. The stress caused by the weight of the hull and the soil layer in the hemispherical ends will be lower by about 50% because these will be rotating through a smaller radius than the cylindrical portion and this means less centrifugal force.

A layer of soil from lunar or asteroid regolith weighing about 11 metric tons per square meter would be about 7 meters (23 ft.) thick if the material has a density of about 1.5 tons per cubic meter. This will be deep enough for tree roots and it will shield the settlement's inhabitants from cosmic rays. One would think that steel 1.22 meters thick or about 9.516 metric tons per square meter would provide lots of radiation shielding. However, it seems that metal will release showers of secondary particles and in the case of an aluminum hull 7.4 cm thick it would actually be worse than no shielding. Results from NASA's On-Line Tool for Assessment of Radiation In Space (OLTARIS) are shown in the table below. https://oltaris.nasa.gov/

| | | | |
|---|---|---|---|
| zero | | | 127.6 mGy/yr |
| Iron | 1.22m (48 in., 9.52 metric ton/m$^2$) | | 54.9 mGy/yr. |
| iron | 952 g/cm$^2$ | Same as previous | 54.9 mGy/yr. |
| iron | 1.83m (72 in., 14.3t/m$^2$) | | 53.3 mGy/yr. |
| Iron + regolith | 1.22m Fe + 3.3m Reg. (5t/m$^2$) | Solar max 2001 | 13.4 mGy/yr. |
| Iron + regolith | 1.22m Fe + 3.3 Regolith (5t/m$^2$) | Solar min 2010 | 19.4 mGy/yr. |
| Steel + regolith | 1.22m steel + 7m regolith (11 t/m$^2$) | | 0.08039 mG/yr. |
| Iron + regolith | 1.83 m Fe + 7m regolith (11 t/m$^2$) | Seems too low | 0.008 mGy/yr. |
| regolith | 7m (11 t/m$^2$) | | 4.32 mGy/yr. |
| regolith | 7m | Nov. 1960 SPE | 0.0326 mGy |
| aluminum | 20g/cm$^2$ (200 kg./m$^2$) | Higher than zero shield | 134.7 mGy/yr. |
| aluminum | 270g/cm$^2$ (2.7 t/m$^2$) | One meter thick Al | 61.7 mGy/yr. |
| Iron + water | 1.22m Fe + 7m H$_2$O | | 0.71 mGy/yr. |
| Iron + water | 1.83m Fe + 5m H$_2$O | | 0.023 mGy/yr. |

GCR 1AU free space, solar minimum 2010, BO2014, unless otherwise noted

# Collision Avoidance in Space

The chances of colliding with another spacecraft, satellite or old upper stage in planetary orbit are low. The chances of colliding with an asteroid or being hit by a meteor in interplanetary space are also low. However, they are not non-existent. Nobody was worried very much about the chances of the "un-sinkable" ship *Titanic* crashing into an iceberg. They were so lax that the builders of the great ship didn't include enough lifeboats. Only fools would not try to avoid a repeat of this disaster in space. Someday, there will be spaceships carrying hundreds even thousands of passengers and valuable cargoes including precious possessions of the wealthy. It would be awful if a luxury space liner had to collide with a space rock and suffer hundreds, even thousands, of deaths just to wake people up to this danger of space travel and get them to do something about it.

Radar systems that can scan a spherical volume of space with a radius of several thousand miles surrounding the ship are called for. Synthetic aperture radar (SAR) might do the job. As of 2010, airborne systems provide resolutions of about 10 cm, ultra-wideband systems provide resolutions of a few millimeters, and experimental terahertz SAR has provided sub-millimeter resolution in the laboratory.[22] Detecting and avoiding even small rocks will be necessary. What if a small meteor broke a window and rapid depressurization resulted? What if a somewhat larger meteor crashed into the ship's nuclear reactor and crippled the propulsion system?

Ships could travel in convoys with enough extra space onboard to take on passengers from a crippled ship. Since launch windows to and from other worlds are often brief in time and months if not years apart, it would make sense for convoys to depart at roughly the same time. Lifeboats won't help much. I seriously doubt that small life boats could carry enough oxygen and other supplies needed to sustain people for weeks or months in space. Some kind of inflatable "escape pods" might be helpful for passengers abandoning ship and moving to another ship in the convoy. That should only take a few hours. Even so, meteor impacts and collisions with asteroids or old satellites should be avoided.

Synthetic aperture radars could scan for objects of all sizes and determine their size and potential danger. They might also use Doppler data to determine the object's velocity along with other radar data to determine distance, position and trajectory of the meteor or piece of space junk and feed that data into computers. If the computers determine that a collision is imminent they could fire thrusters or orbital maneuvering system rockets to get the ship out of the way without excessively altering the ship's course in a way that would cause it to miss its destination by thousands of miles or less. The computers would also alert the crew but they wouldn't wait for slow humans to redirect the ship. There could also be radar buoys in space that scan for dangerous objects and transmit data via relay stations with radio or laser to ships in space. A central data bank containing information about the orbits of just about every space rock down to less than 1 cm in diameter and pieces of space junk could be created and updated constantly. This data would be used when calculating a ship's course through space to ward off collisions.

When tens of thousands of ships are plying the space of the solar system knowing their course or trajectory ahead of time when planning a ship's course could also be a lifesaver. Collisions of large ships traveling at high speed, manned or robotic, would probably be fatal to all onboard. The cost of losing a robotic ship with its freight would also hurt someone's bottom line, and that's not desirable. Ships that have no flight plan stored in the central computer data bank will be highly suspicious. Radar buoys might locate them and alert the authorities. Ships would probably be fitted with transponders too. The collision avoidance systems could also prevent smuggling and human trafficking in the future.

The computer power would be immense and the code would be millions of lines long. The mechanical and electronic components of the radar dishes and transceivers would also be impressive. Whoever defeats all these challenges and manufactures a product that could be required by law to be fitted on every spaceship that carries humans could make a lot of money in a future time when millions of people travel in space every year, or every day! Robotic cargo ships might not be required by law to have these systems but the insurance companies might require it. The law might require robotic ships carrying emergency supplies of food or medicines to have collision avoidance radars also.

# Space Food

Many foods were developed for the Space Shuttle program that could also be eaten by future space workers and tourists. Here are some of those foods.

## Natural Form (NF) Foods

Nuts, granola bars, and cookies are natural form foods. They are ready to eat, packaged in flexible pouches, and require no further processing for consumption in flight. Both natural form and intermediate moisture foods are packaged in clear, flexible pouches that are cut open with scissors. In the future, it should be possible to provide some of these foods straight from the supermarket to space.

## Irradiated (I) Meat

Beef steak was the only irradiated product used on Shuttle. Steaks were cooked, packaged in flexible, foil-laminated pouches, and sterilized by exposure to ionizing radiation so they remained stable at ambient temperature.

## Shelf Stable Tortillas

Flour tortillas are a favorite bread item of the Shuttle astronauts. Tortillas provide an easy and acceptable solution to the bread crumb and microgravity handling problem, and were used on Shuttle missions. However, mold was a problem with commercially packaged tortillas, especially on longer missions which had no refrigeration.

A shelf stable tortilla was developed for use on the Shuttle. The tortillas are stabilized by a combination of modified atmosphere packaging, pH (acidity), and water activity. Mold growth is inhibited by removing the oxygen from the package. This is accomplished by packaging in a high-barrier container in a nitrogen atmosphere with an oxygen scavenger. Water activity is reduced to less than 0.90 in the final product by dough formulation. This reduced water activity, along with a lower pH, inhibits growth of pathogenic clostridia, which could be a potential hazard in the anaerobic atmosphere created by the modified atmosphere.[23]

It seems we could use supermarket tortillas if we have refrigerators in space, but that would depend on the size of the ship. The specially packaged tortillas described above might be preferrable in small capsules and life pods. Some other foods enjoyed in space are Swedish meatballs, yogurt, chicken soup, tortillas, shrimp, hot sauce, M&Ms and dried produce.[24]

Eventually, weightless ships and stations will be replaced by ships with centrifuges and rotating stations with "artificial gravity." It will then be possible to cook foods normally on a stove top or in a conventional electric or microwave oven. For instance, it will be possible to boil pasta and pan fry things. Vegetables could be chopped up on galley cutting boards. Foods that crumble like corn bread could be eaten. Many conventional supermarket foods like pasta, wheat flour, corn flour, powdered eggs, powdered milk, pancake and waffle mix, potatoes, dried fruit, nuts, and more could be shipped into space for tasty home cooked meals. They would be hydrated with recycled water. Some canned foods might be shipped up too. Refrigeration will be desirable in larger ships and stations. Foods from Earth would be very expensive even at the rock-bottom price of $67 per kilogram to LEO. In time, food must be produced in space at large stations and on the surfaces of other worlds in pressurized habitat and this offers many challenges.

We should be able to grow all varieties of vegetables, fruits, beans, nuts, grains, spices and mushrooms in vertical farms or stacks of trays with UV filtered sunlight delivered by fiber optic light pipes. During darkness, highly efficient red and blue LEDs could be used. Mushrooms are rich in protein, don't need light and can be grown on wastes. The farms might use hydroponics with nutrient solution derived from wastes. In large stations like Kalpana Two and on the Moon there could be some "outdoor" shrubs in various living spaces and other decorative plants cultivated in soil as long as soil bacteria can be controlled. Dwarf fruit trees might also be grown for fresh cherries, oranges, raspberries, apples, peaches, figs, peaches, apricots, lemons, pears and avocados. These would also please the eye.

Cattle, sheep, goats and pigs might be too much trouble. Sure, we could ship sedated tube fed diapered calves and piglets to the Moon but when they mature they will require lots of water and feed and excrete tremendous amounts of manure. It only takes a few days to get to the Moon but I doubt sedated animals will do well on a six month voyage to Mars or even a 39 day voyage. Martians would also have trouble creating shelter for these beasts, growing feed and dealing with all the manure, so the Martians are not better off than Lunans in this regard.

This means no bacon or steak unless it is imported for several hundred dollars a pound and we will have to drink soymilk. Soymilk is okay and healthy too. What about butter and cheese? We could make margarine from vegetable oil and we might get used to tofu and soybean curds.

According to Bryce Meyer, fish and shrimp could be cultivated at early settlements, but chickens and turkeys would require larger populations to create a self sustaining biosphere.[25] When our habitations grow large enough, chickens and turkeys could be kept so we can eat Thanksgiving and chicken dinners often. We could just rocket up fertilized eggs and hatch them in incubators. Fish, shrimp, lobster, crabs and clams could be kept in water tanks also. I suppose we could transport their eggs to the Moon or Mars and hatch them there.

Some people will eat rabbit, guinea pig (cavies), frog, turtle, snails and insects like grasshoppers, mealworms and chocolate covered ants. It should be fairly easy to keep these small animals and bugs. So these can supply nutritious animal protein too.

Cats and dogs could be fed with chicken and turkey entrails, rabbit and guinea pig guts, etc. So we should be able to keep pets. Carnivores love to eat the guts.

If we want beef we will have to grow it in culture dishes should that become possible at a reasonable price. It might be possible to keep just a few cows and have a rare sacrifice and a beef feast; perhaps hamburgers for a big sporting event or circus. Space people aren't going to be eating lots of bacon cheeseburgers on the Moon or Mars, so the fast food industry will have to adapt and sell fish, chicken and turkey sandwiches, rabbit and bug nuggets, shrimp etc.

Turkey bacon isn't so great because it is so dry but spaghetti and ground turkey is good enough....no parmesan cheese though unless tasty plant derived cheese substitutes are developed.  This calls into question the wisdom of sacrificing a cow that can produce real milk for cheese and other foods like cream and butter.  Perhaps sports fans should forget about burgers and dogs and eat ice cream, baked potatoes with cheddar cheese and sour cream, and grilled cheese sandwiches!

Space food will be different and all kinds of new dishes might be invented or rediscovered. The menu will be full of chicken, turkey, fish and other small animal dishes. Beef and mutton eaters will just have to live with it or go somewhere else.  Many recipes are not heavy on meat and use it mainly for flavoring. Shredded chicken in rice with chickpeas, tomato sauce and curry is good stuff and doesn't require lots of meat.

Algae grows like crazy and during lunar dayspan with UV filtered sunlight it could be possible to grow tons of it and possibly use it for livestock feed all during the night and then some.  It is rich in protein, oil and cellulose. Using algae along with leaves and stems from crops for fiber in livestock diets could make it possible to produce feed without using much farm space.

With chickens there could be real eggs for mayonnaise. Hollandaise sauce from eggs, cream (from soy?), lemon juice and a little red cayenne pepper might be made.  So it might be possible to make a good turkey bacon BLT sandwich with  mayo (or hollandaise) to moisten up that dry turkey bacon. There could be ketchup and mustard too which just about everyone likes. So there could be turkey burgers with ketchup, mustard, onions and pickles at off-Earth McDonald's. Will the fries be made in vegetable oil or will air frying be the future?

# Orbital Clean-up

Riding external tanks to low Earth orbit might seem to some critics as just another way to litter LEO with space debris and create a junk yard in space. This is unlikely. The tanks would be used to build space stations and other structures like propellant depots and they might even be melted and sprayed into powders for rocket fuel. Solar powered conductive tethers could be used to move the tanks around without expending any propellant. When direct current is fed into a conductive tether, it exerts a Lorentz force against the geomagnetic field and the tether exerts a force on the spacecraft that either accelerates or decelerates the spacecraft. By accelerating the vehicle its orbit can be raised and by decelerating it the orbit can be lowered.[26] By changing orbits spacecraft can rendezvous with each other. If numerous spacecraft with solar panels and tethers are orbited, they could clean up the dead satellites and spent upper stages in LEO without using any propellant that would be so costly to supply. Unfortunately, there is a lot of space junk in GEO or in graveyard orbits at higher altitude where the Earth's magnetic field is not strong enough for electrodynamic tethers to operate.

Resources from the Moon could solve this problem. Hydrogen from lunar polar ice, sodium or magnesium might be used as reaction mass in highly efficient solar electric rocket propelled spacecraft that grapple with space junk and haul it to a high orbital space station where it is recycled. Thousands of tons of aluminum, silicon, composites, gallium and other materials including precious metals are floating around up there and going to waste. They only serve as navigation hazards. There is certain to be some profit in the salvaging of this space debris if propellant is cheap enough. Electrodynamic tethers might not work in and above GEO but solar and magnetic sails that don't use any propellant at all might also be applied.

# Lunar Railroads

Science fiction stories often picture magnetic levitation (Mag-Lev) trains on the Moon hurtling through the vacuum as fast as jet airliners. This might be real in the more distant future, but with Mag-Lev costing about $100 million per mile it seems more plausible to predict slow freight trains on conventional rails in the near term.[27] High speed driving for trucks and other ground vehicles over rocky dusty lunar terrain or dirt roads seems unlikely, but not impossible. Railways seem to be the best candidates for mass transport on the Moon. If a mining and mass driver launcher base is built on the equator at 33.1 degrees East, a rail line to an ice mining base in the South Polar region would be very helpful. Tank cars full of water, liquefied methane and carbon oxides, and ammonia from ice could haul the stuff to the mass driver base. The liquids would be filled into small aluminum or iron shells with a loaded mass of about 40 kg. each. These would be fired into space and caught by mass catchers at L2 then hauled down to GEO and LEO.

Rails could be produced by rolling iron or steel. While an electron beam additive manufacturing device could lay down 7 to 25 pounds of metal per hour, heavy rails could be rolled in minutes, even seconds.[28] European rails weigh about 40 kg. to 60 kg. per meter.[29] With 80 tons of carbon from volatiles mining, 24,000 tons of mild steel could be made. This would be enough for 240 kilometers (about 150 miles) of track. Much more than this would be needed since the distance from the lunar equator to the South Pole is about 1,700 miles. About 274,000 metric tons of iron or steel would be called for. A similar amount of metal will be needed for the flat plate ties.

The choice of metal is important. Wrought iron was used for rails in the 19th century before the Bessemer process made large quantities of mild steel available. Wrought iron and pure iron extracted from regolith have similar properties. Both have about 40,000 psi compressive and tensile strength. Pure iron is softer than steel, but it might work given the low gravity and lower pressure on the rails. The choice will depend upon the availability of steel and its cost. Wheels, undercarriages and springs would probably require alloy steel with added chromium, nickel and/or manganese. Lightweight titanium for these parts is also possible. Cars including tanks could be made of lightweight aluminum.

There will be no rust on the Moon but high dayspan temperatures will require gaps in the rails to allow for expansion and solar shields made of foil or sheet metal. In the super cold of nightspan the rails could become brittle and crack. It may be impossible to run at night unless electric heating elements drawing power from the third rail or overhead power cable are installed to keep the rails warm.

The rails and ties would be made at the equatorial base and robotic vehicles would ride the tracks to their end and lay down some more track. They would then have to ride back to base to pick up some more rails and ties. This might not be efficient. Perhaps wheeled vehicles that ride over regolith will cycle between base and track laying sites in a constant loop. They will get power from a third rail or overhead cable which will require towers to support it. Power would come from a system of three solar power plants built on the lunar equator to supply electricity 24/7 to the mining base, its mass driver and the railway.

An overhead cable could be made of aluminum or possibly calcium clad in aluminum. A third rail could be made of aluminum or calcium clad in iron or steel, since iron is a poor conductor. Solar panels will generate D.C. This would have to be inverted to A.C., stepped up to high voltage with transformers and fed as is into the overhead cable or third rail or rectified back to high voltage D.C. To transmit power over a thousand miles extremely high voltages will be necessary.

The metals, rails, ties, cars, solar panels and other parts of the system would all be produced on the Moon. Robot labor would be used extensively. The only alternative to a railway would be a pipeline from polar ice mining bases to the equatorial mining and launching base. A pipeline would face the same problems caused by temperature extremes. However, a pipeline can't have gaps but it might be protected from thermal extremes by foil or sheet metal solar shields and have heaters. Naturally, a pipeline cannot move any cargo like machines, habitat and such that a train can. A railway to supply necessary equipment to the polar bases is superior. The space stations in GEO and LEO cannot expect hydrogen, carbon or nitrogen from the Moon in large quantities until a circum-lunar equatorial power grid and railways connecting the bases are built. Until then, the only way to get HCN in Earth orbit is to "piggy-back" it with other payloads rocketed up from Earth.

This would be a major project. Since mare regolith is about 12% iron, about 2.3 million metric tons of regolith must be processed to get 274,000 metric tons of iron. Excavating is not impossible. Mare regolith has a density of about 1.5, therefore a square kilometer dug to a depth 1.53 meters would supply this much regolith. The challenge would be constructing machines like Dr. Peter Schubert's supersonic dust roaster and all isotope separator (SDR-AIS). Some parts, like the thorium oxide free fall tubes would have to be imported. The rest could be made on the Moon. Another challenge is generating enough power to operate the SDR-AIS systems as well as all the other machines at the base. Work would have to start out small and grow big. Solar panels would be one of the first things to produce and manufacturing capacity for solar panels would have to be expanded as the demand for power grows. Silicon doped with aluminum or phosphorus solar panels would be made since elements like gallium, arsenic, germanium and indium are so rare on the Moon. Radiation and heat will take their toll on the panels. They would have to be replaced every eight years or so. Such is the price of doing business on the Moon, assuming it is cheaper to produce silicon solar panels instead of importing gallium based multi-junction solar panels.

Railroad building would not stop with the 33.1 degree equatorial base to southern polar ice mining base line. Railroads could stretch across the mare and serve numerous bases on mare-highland "coasts." Some of these later bases would be mining sites and others would be resort hotel towns. Bridges across rille valleys might be built with cast basalt blocks made at nearby sites in the mare. Railways could extend to the North Polar regions and to the far side of the Moon where giant radio telescopes are built away from the Earth's radio noise. This would be part of the search for extra-terrestrial intelligence.

Rail travel and shipping will not be free. Somebody is going to have to spend a lot of money to build these things. That individual or corporate entity will have to charge customers whatever it takes to make a profit before the bonds mature and investors want their take. Solar power satellite building and lunar tourism could be the two major drivers of lunar industrialization starting with a small "seed" and bootstrapping base to supply raw materials including liquid oxygen to companies working in GEO and LEO.  The demand for light elements like hydrogen, carbon and nitrogen from ice for rocket fuel and life support will spur the development of lunar railroads.  The creation of a space faring civilization will not end with government and taxpayer financed projects.  It will be paid for with million dollar checks written by wealthy people who want out of this world adventure.

# Miscellaneous Products in Space

Space settlement presents some odd challenges that may seem like impertinent details; but won't when you have to use the toilet. There are no trees on the Moon and if there are, they are likely to be dwarf fruit trees for food. In sealed and pressurized lava tubes trees could be grown for decorative purposes but it isn't likely that there will be enough of them to make much paper. The same situation will exist in orbital or free space settlements, on Mars and any other worlds we inhabit in this solar system. Paper will have to come from hemp, straw, rice hulls, peanut shells and anything with vegetable fiber. Offices could be almost paperless and put all written documents in computers. Cash could be replaced by debit cards, credit cards and smart phones. Banks could all be computerized. Bills could all be paid online. Paper napkins, paper towels and facial tissue could all be replaced by machine washable cloth items. Paper towels in rest rooms could all be replaced by hot air driers. Paper bags could be replaced by reusable basalt fiber sacks. Cardboard mailing packets could be replaced by basalt fiber envelopes and basalt fiber packing materials. There could be reusable metal mailing boxes too. Stamps and labels may demand some paper from the hemp farms and agricultural wastes. Toilet paper becomes the last refuge of the paper merchant. Hemp and agricultural wastes will be allowed for toilet paper first; stamps and labels second. Stamps and labels might even be replaced by programmable electronic chips and chip readers. Without toilet paper, or bath tissue to be more polite, there would only be bidets and personal cleansing cloths that are machine washed in very hot water with a little bleach after use.

There is no oil on the Moon. Unless life once existed on Mars there will be no oil on Mars. There are hydrocarbons in asteroids that take the form of a tarry substance that resembles kerogen. Light elements like hydrogen, carbon and nitrogen exist in the ices of outer moons and the atmosphere and methane lakes of Titan. When these latter resources are tapped it will be possible to make large quantities of plastics, silicones, lubricants, elastomers and other items made from oil today on Earth. Even so, these products will all be reused, repurposed or recycled. It's not likely that asteroid mining, ice mining or mining on Titan will be dirt cheap.

The Moon does contain hydrogen, carbon and nitrogen from the solar wind implanted in the regolith. Mining millions of tons of regolith and heating it to about 700 C. will release these elements. The hydrogen and helium will outgas as is and some hydrogen will react with oxides to form water. Carbon will react with regolith oxides to form CO and $CO_2$ and it will react with hydrogen to form methane. Nitrogen will also off gas. At higher temperatures sulfur, sodium and potassium will volatilize in the vacuum.

The only other source of these light elements are lunar polar ices. At present, not much is known about the ice. It is not known if the ice exists in thick sheets that would be harder than steel at supercold temperatures or as crystals mixed in with the crater floor regolith that could be extracted electrostatically or by heating with waste heat from small nuclear reactors that power mining machines. Billions of tons of ice are suspected to exist. This precious resource will require a great deal of work to tap into and it must not be wasted. However, the ice will have many uses from making silane rocket fuel to drinking water.

Carbon dioxide can be subjected to electrolysis to get carbon monoxide and oxygen. Carbon monoxide can be reacted with hydrogen to make almost any organic chemical given the right temperatures, pressures and catalysts. On Mars, carbon dioxide is abundant in the atmosphere and water for hydrogen is present in the soil. This is one of the reasons Mars is richer than the Moon and a better target for settlement in the eyes of some. Even so, the Moon's solar wind implanted volatiles and ice can make Closed Ecological Life Support Systems (CELSS) possible as well as silane rocket fuel production and chemicals for synthetic materials. These lunar resources would only have to be exploited until asteroid mining is in full swing several decades after Moon mining has begun. The Moon's precious light elements could be used to launch space industry but they wouldn't have to be savagely wasted over the long run.

Processing mixtures of hydrogen and carbon monoxide, also known as synthesis gas, is not the only way to make synthetic materials. Many substances, like Poly-Lactic Acid (PLA), can be made biologically. Corn can be grown to make compostable PLA. Composting is an easy way to recycle. Plants can take $CO_2$, water and soil minerals and make all sorts of things from cloth fiber to drugs, lubricants, dyes, and other products. Plants are the ultimate chemical factories and they can be erected from nanomachines-seeds! Seeds are Nature's nanomachines. Natural and genetically modified yeast, fungi, algae and bacteria can also be put to work. Cultures of yeast, algae, bacteria and fungal spores are also natural nanomachines. Compact, lightweight and easy to transport, we just need water, air and growth media within pressurized habitat, made perhaps by 3D printing with molten basalt, to get these natural nanobots working for us.

Cotton and hemp could be cultivated along with food crops to make clothing-an essential. With thread, yarn and fabric, tailors and people who like to knit or crochet in their off-hours can make all sorts of clothing items. Fabric will also be needed for towels, sheets, bandages, etc.

Woven basalt fiber fabric can be used for furniture upholstery and cushion stuffing. Mattresses could be made, but cotton sheets would be desired. Basalt fabric is non-toxic and smooth, unlike scratchy glass fibers, but it doesn't have wicking ability. Moisture wicking might not be important for furniture but it will be for underwear, pajamas, sheets and towels.

Large animals like sheep for wool are out for a while until larger populations exist, if I understand the space farm experts correctly.[30] Cows for leather are even less likely. There are alternatives. Bolt Threads makes a leather substitute from fungal mycelium and spider silk threads from genetically programmed yeast. They will be offering numerous products beginning in 2021.[31] This will not only help protect the Earth, it will be ideal in space.

It will be up to the space farm experts to figure out how cotton and hemp will effect mass balances in the CELSS. Cloth will take carbon and hydrogen out of the loop and sink it into various sundry items. These could be ripped and recycled when they get old and worn. When they can't even be recycled or reused they can be composted or subjected to biological degradation in yeast and bacteria filled bioreactors. Paper items will meet the same fate. Paper can only be recycled a given number of times depending on what it is made of like trees or hemp, and toilet paper goes straight into the compost or bioreactors.

The semi-synthetic fiber Rayon is a possibility. It should be possible to produce sodium hydroxide, carbon disulfide and sulfuric acid needed for Rayon production on the Moon, but these are nasty chemicals. Recycling is necessary and some nasty byproducts form like highly toxic hydrogen sulfide.[32] It would be wise to do this processing in modules separate from the main base module complex.  A source of cellulose is required.  Bamboo, cotton, hemp and algae are candidates for cellulose. Bamboo grows fast but algae might be much more productive.  Rayon is smooth but it is not very strong or machine washable. Items of Rayon would have to be dry cleaned.  It's also very flammable but there isn't likely to be much cigarette or cigar smoking in outer space and stoves will all be electric.  Some people might prefer to avoid Rayon.

Nylon and polyester involve some complex chemistry and chemicals from oil not available on the Moon.  We could probably make the requisite chemicals from hydrogen, carbon and nitrogen from regolith, ice or asteroids.  Polyester requires ethylene glycol and ethylene is makeable on the Moon if hydrogen and carbon monoxide are produced. [33]

Given the nastiness of making some things from chemicals we must either use robots in isolated modules or rely on biological sources.  Would anybody think we are crazy for having real silk worms??  There is a market trend towards natural substances that's been going on for decades.  Natural, no pesticides/herbicides, organic locally sourced foods and such are popular with people who can afford them!!!

Furniture making with basalt fabric upholstery and basalt fiber cushion stuffing has been mentioned.  How will furniture be made without wood or even structural plastic? Frames could be made of metal tubes.  Basalt bricks and basalt slabs could be used for bookcases, tables, benches, countertops and probably other things.  Autoclaved Aereated Concrete (AAC) could also be used.  This material is made from sand, cement, water and aluminum powder.  The ingredients are mixed together and heated in an autoclave.  The aluminum powder reacts with water and forms hydrogen bubbles that give the finished substance its porous structure and extreme lightweight.  It is very strong, yet it can be cut with a saw, nailed and drilled with ease.  It's the closest thing to a wood substitute to be found on the Moon or elsewhere in space. Of course, it could not be used to make paper.  Furniture designers could create all sorts of interesting furniture with AAC.  Coating it with paint is possible.  Paint would have to be made from chemical or biological sources.

# Space Liners & Centrifuges

Artificial gravity plating is about as unlikely in the near future as negative matter reactionless drives. The only way to provide a substitute for gravity is to use centrifuges. The first spaceships will be weightless. They will be arranged sort of like jet airliners, use chemical rocket propulsion and reach the Moon in about 30 hours. That might seem like a long time to sit in a couch even if the seat can recline and you can sleep. Since passengers will be weightless they won't get very sore sitting for a long time but they will want to move around and stretch their limbs. Weightless floating rooms with observation domes perhaps will be provided to make the flight more enjoyable and increase passenger comfort.

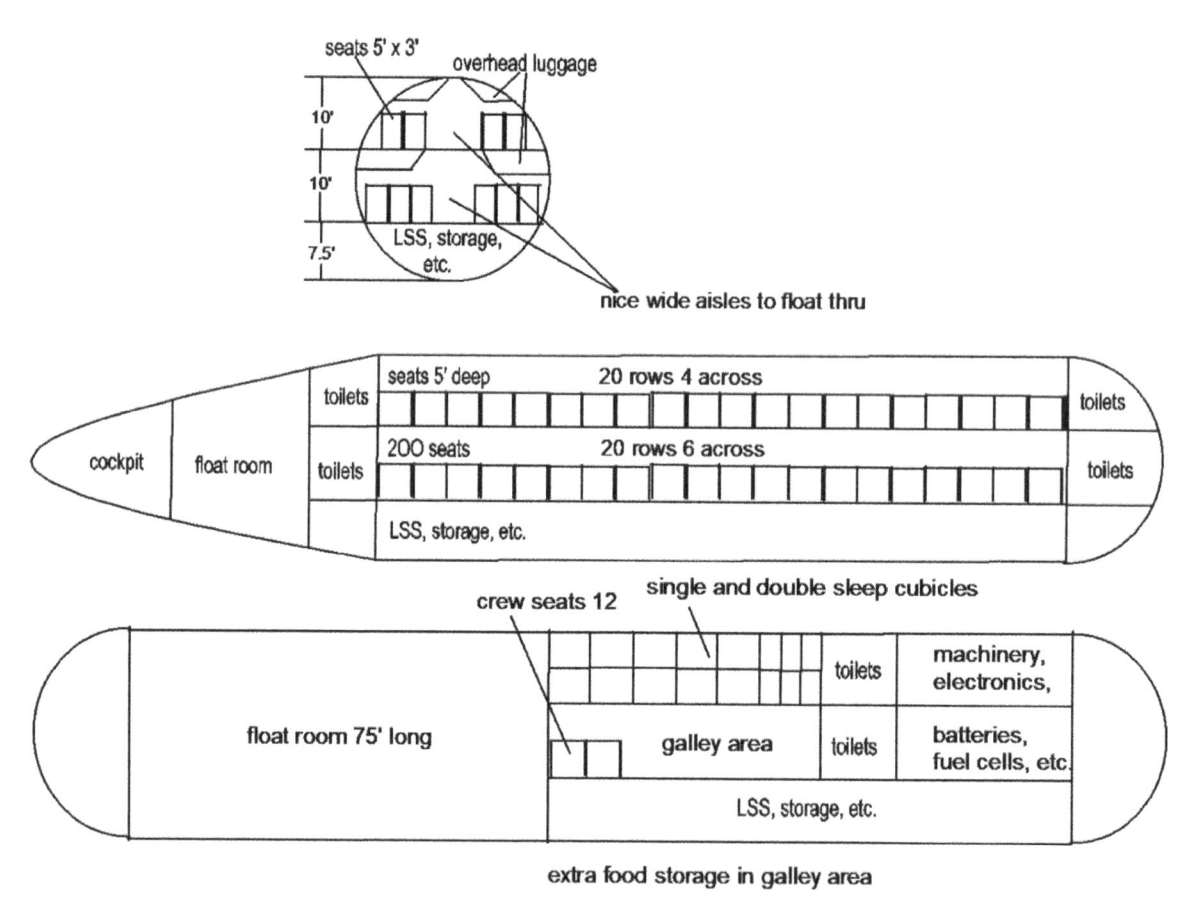

Fig. 9 Passenger seating arrangement on weightless ship with floating rooms.

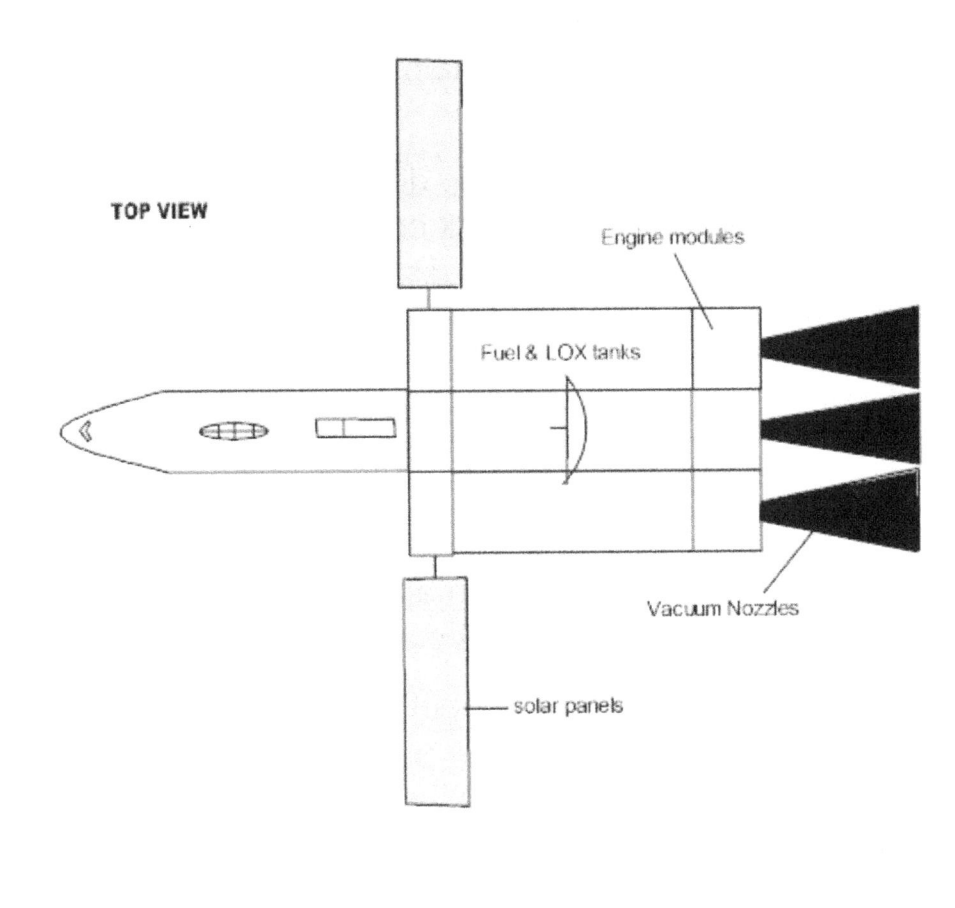

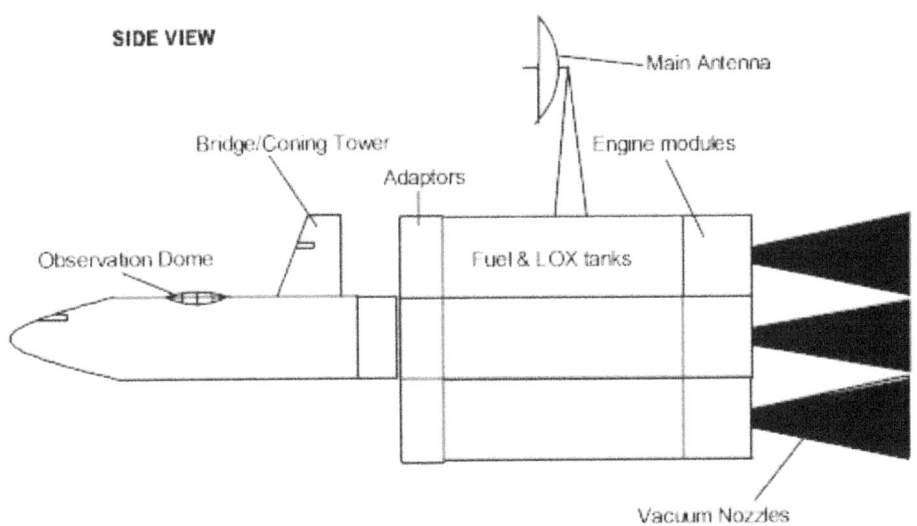

CHEMICALLY PROPELLED INTERLUNAR PASSENGER SHIP 30 HRS. LEO TO EML1 OR LLO

Fig. 10 External view of weightless interlunar spaceship.

If the ship operates on the same time schedule as the launch base and everyone has adjusted to that time, lights could be turned down and everyone can sleep at the same time. If not, then there could be curtains around seats and if that doesn't help there could be sleeping cubicles as pictured above in the rear hull.

There will probably be people who dislike this arrangement, perhaps because they get really space sick or want more privacy, who will prefer to travel in private cabins in centrifuges that provide "artificial gravity." While complete spaceship hulls could be built on the ground and rocketed into orbit where they are docked together, centrifuges will have to be built in space from metals obtained from spent external tanks and possibly lunar materials. Ships with centrifuges will be later generation vehicles. These centrifuges will add mass to the ship so it might take more like 80 hours to reach the Moon in such ships. That means the ticket price must be higher because the ship won't make as many round trips per year. Equipping the ships with centrifuges will make them more expensive also.

The centrifuges will be about 90 feet in diameter and rotate at 3 RPM to produce just a little less than lunar gravity equivalent. The cabins will be about 14 feet by 14 feet including the private toilet closet. Water for flushing will be stored in tanks and pumped through lightweight plastic pipes to the low mass compact plastic toilets. The sewage will be pumped through plastic pipes and into an electric distillation machine in the centrifuge. The distilled water will be filtered, treated with chlorine or ozone and reused for flushing toilets.

The centrifuges will rotate in opposite directions to cancel out gyroscopic effects. Flywheels on the ship's roll axis might also be needed to control the ship's attitude as angular momentum changes when people move in and out of the centrifuges. There will be centrifuges with private cabins that are entered by ladder thru a sliding door in the ceiling instead of wasting space with wide corridors. There will also be centrifuges with a galley where meals can be cooked normally thanks to the "artificial gravity" and passengers can sit down for a meal, pour drinks, etc. There will also be exercise rooms and hot showers. The water for the showers might be recycled after distillation, filtration and chemical treatment.

CHEMICALLY PROPELLED INTERLUNAR PASSENGER SHIP 72 HRS. LEO TO EML1 OR LLO

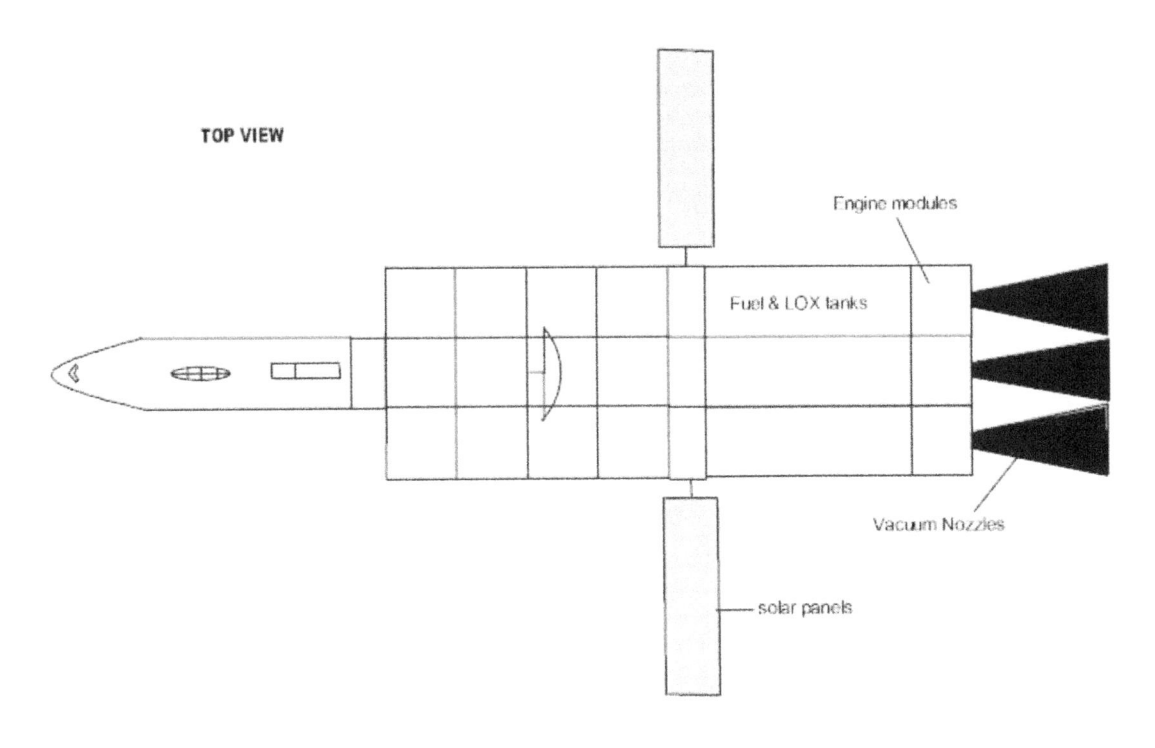

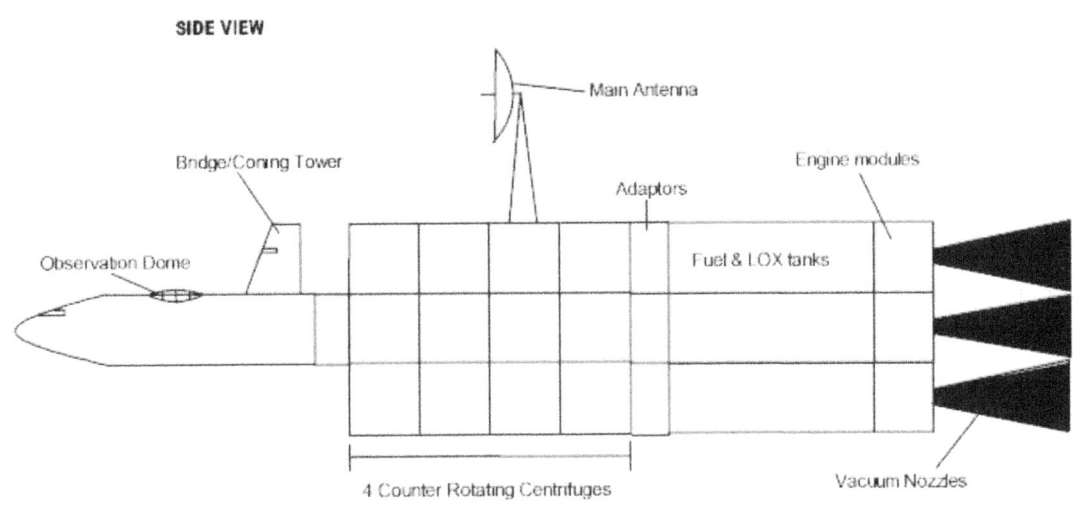

Fig. 11  Ship with centrifuges

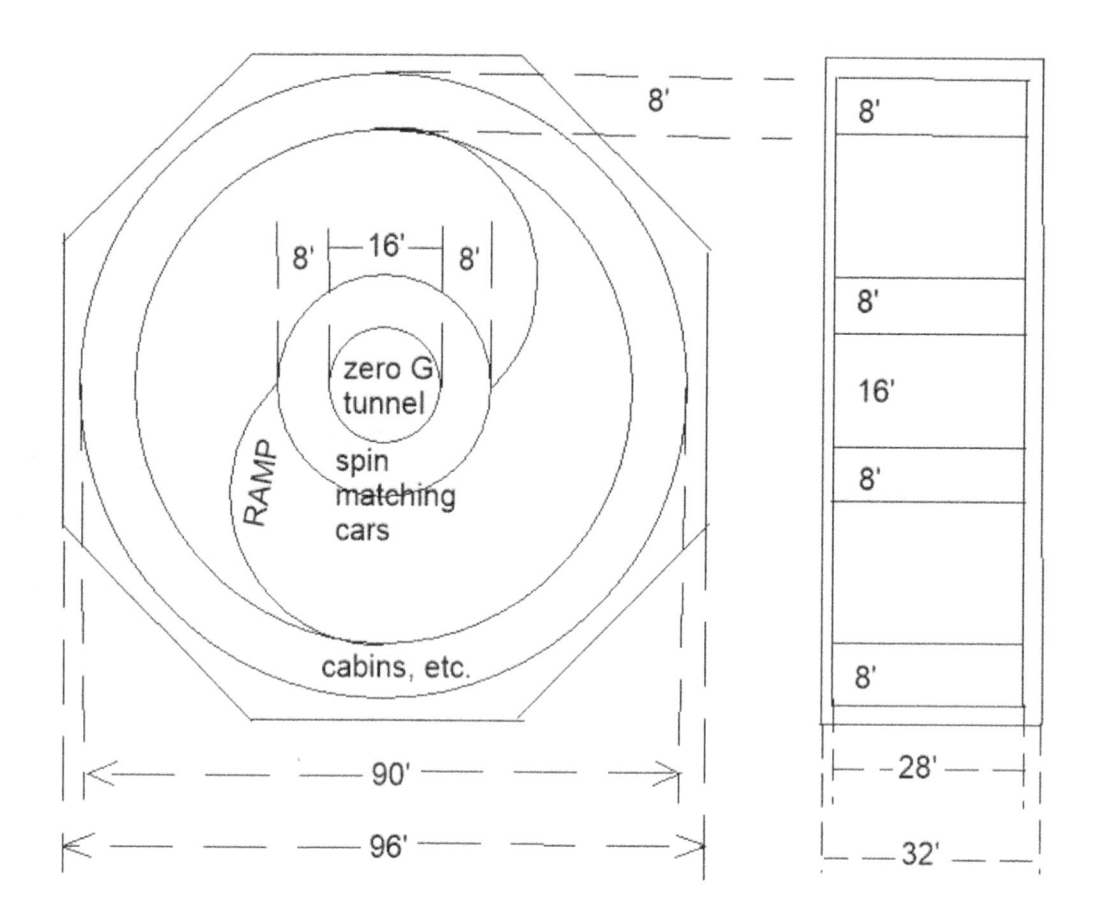

GENERAL CENTRIFUGE LAYOUT

Fig. 12 General layout with dimensions

Passengers will float through the central tunnel, possibly pulling themselves hand over hand along ropes. Getting around in weightlessness will require some minimal physical ability. They will then enter some stationary "elevator" cars that when filled with people will rotate until matching the spin of the centrifuge. Then they can walk down ramps to the cabins or mess hall in the rim.

Moving in and out of centrifuges, that is transferring from low G to weightlessness then low G again might make some people ill. Even with centrifuges it won't be possible to escape from weightlessness. Of course, weightlessness is the whole reason many people want to travel in space! Medications to control motion sickness may be necessary for some people. Low gravity will make life onboard more natural and more comfortable, but not much is known about the long term effects of low gravity. It is known that prolonged weightlessness is very unhealthy. Ships to Mars and other destinations in the solar system will need centrifuges.

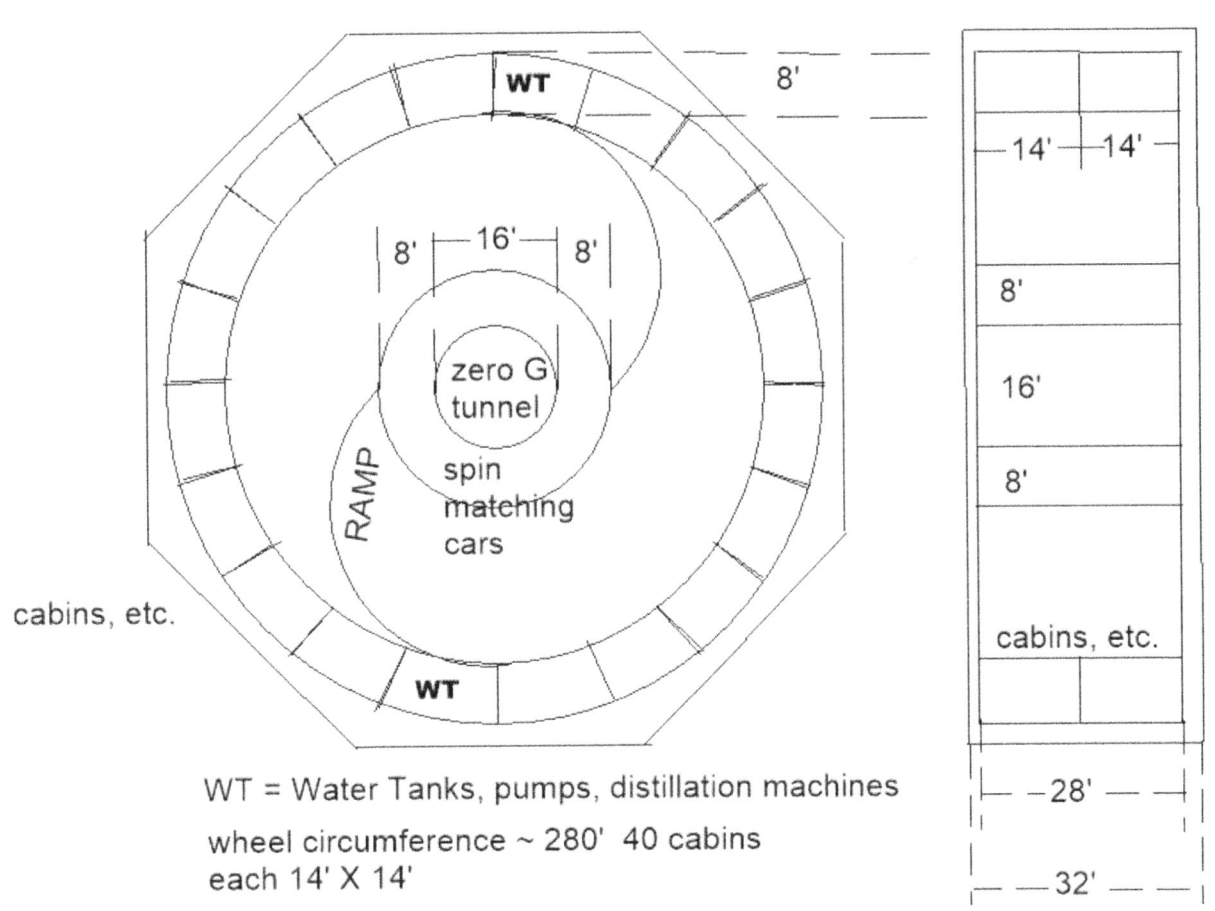

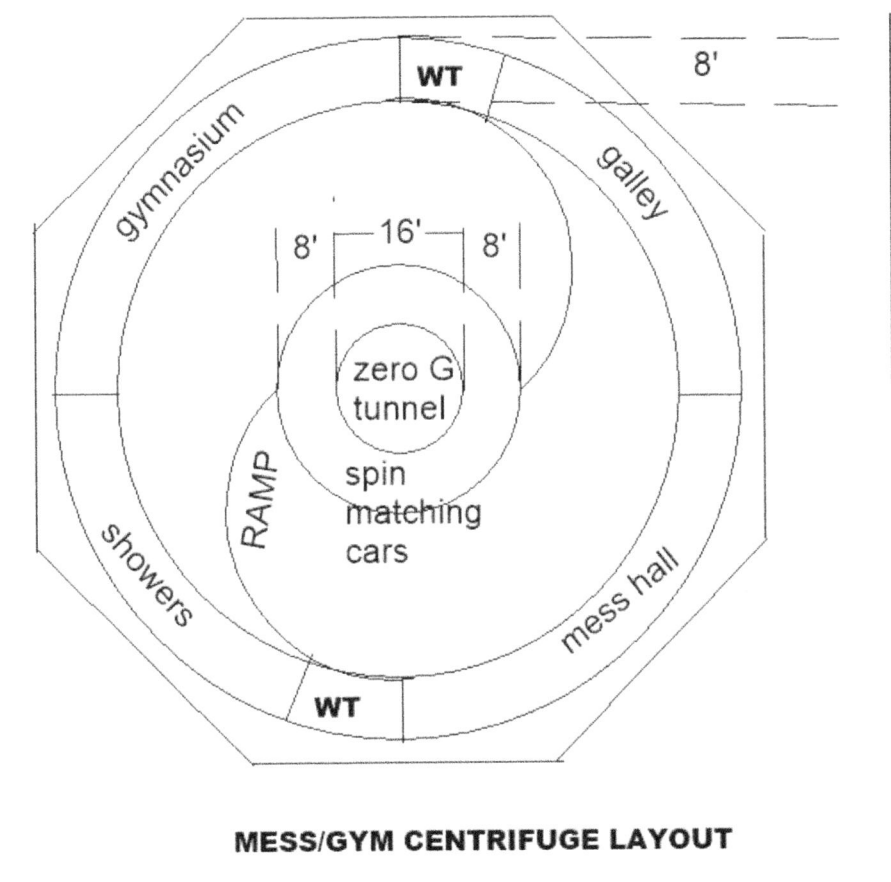

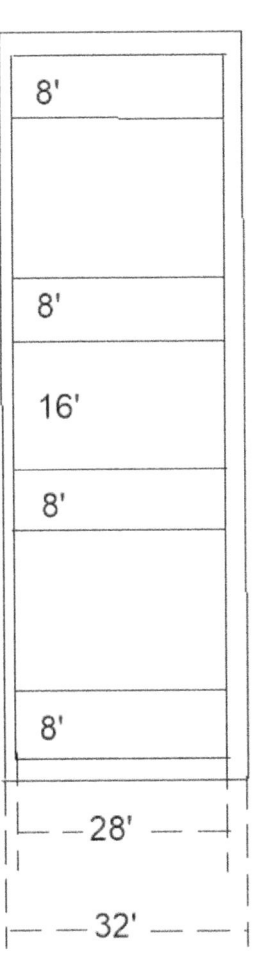

WT = Water Tanks, pumps, distillation machines
wheel circumference ~ 280'  40 cabins
each 14' X 14'

**MESS/GYM CENTRIFUGE LAYOUT**

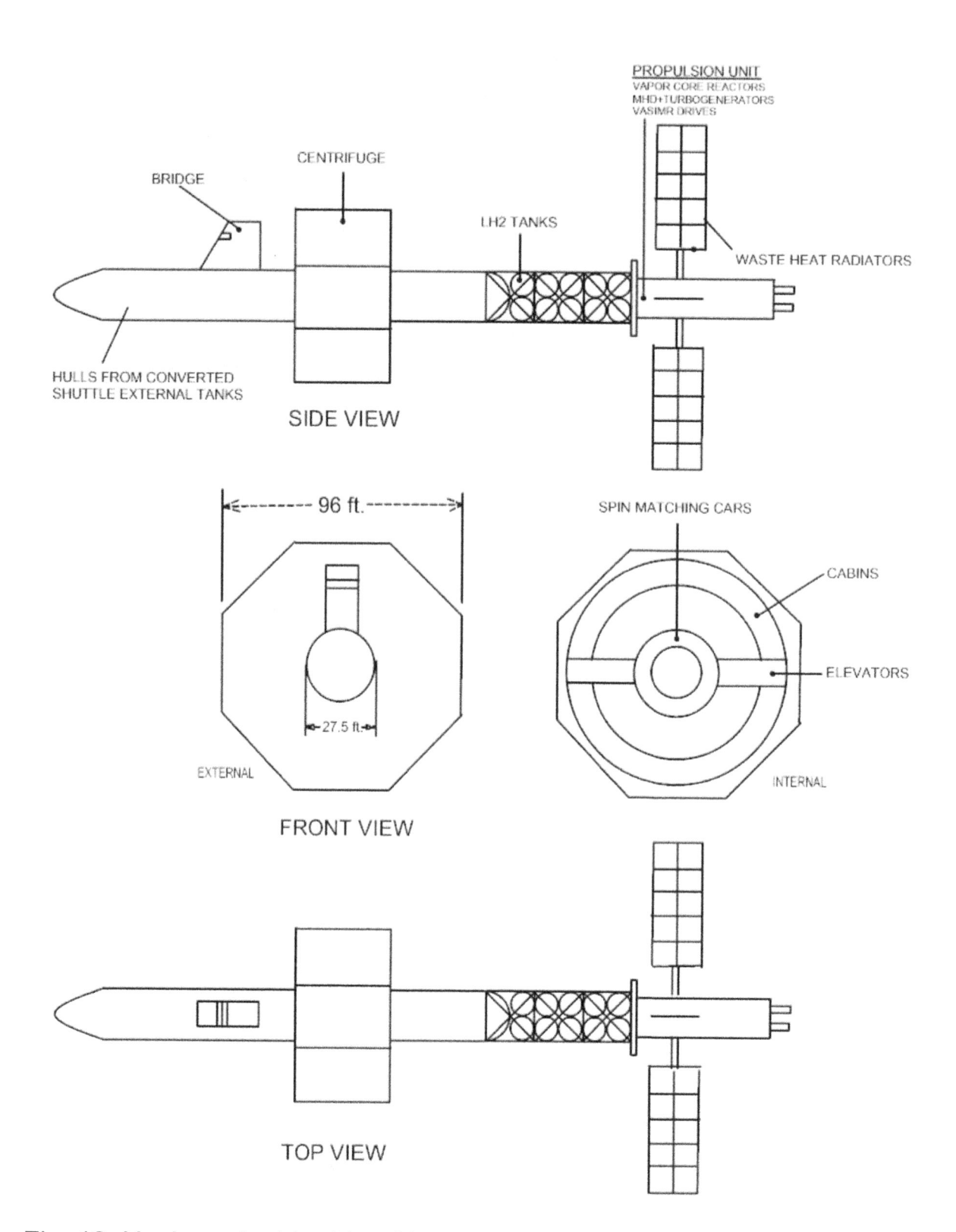

Fig. 13  Nuclear electric ship with one centrifuge countered by roll axis flywheels.

TWO COUNTER-ROTATING PASSENGER CENTRIFUGES

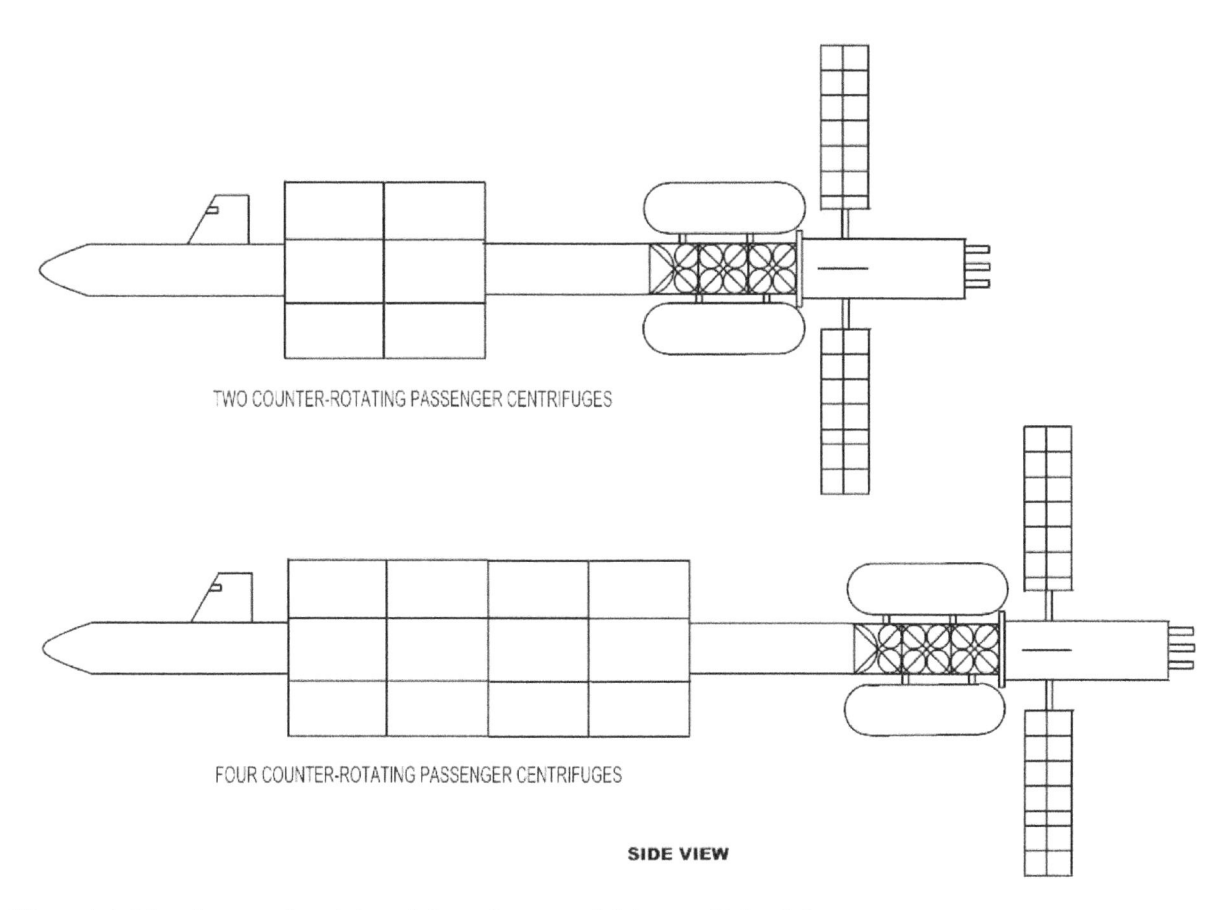

FOUR COUNTER-ROTATING PASSENGER CENTRIFUGES

SIDE VIEW

Fig. 14 Nuclear electric ships for rapid transit to Mars.

Interlunar ships could use chemical propulsion. Lunar liquid oxygen and silane mixed with metal powders could be used for propellants. This will not frighten those who are worried about nuclear reactors in orbit or nuclear ships retro-rocketing into LEO on their way back from the Moon. Governments will probably restrict space nuclear power. Hopefully, the law will permit nuclear ships in GEO. Chemical propulsion could get settlers to Mars in six months but nuclear electric propulsion could cut that down to 39 days. Marsbound travelers could ride space taxis from LEO to GEO where they board nuclear ships to Mars.

No hard evidence exists to back this claim, but 39 days in the lunar equivalent gravity of the centrifuges probably has no severe health consequences. As for travel beyond Mars, larger ships with larger centrifuges producing higher G force and active magnetic cosmic ray shields are called for.  Someday, these ships will be built in space from materials mined in space. Their nuclear fuel will also be mined in space from the Moon and asteroids.  They might use nuclear fission reactors and electric propulsion, or they might have helium 3 burning fusion drives.  It could take many months even years to reach the outer solar system.  A minimal amount of gravity will be needed to stay healthy on long space voyages.  It might be necessary to produce 1 G and it might be possible to live at a reduced level of say 0.4 to 0.7 G. Nobody knows presently.

What will these ships cost?  We can guess several hundred million dollars in today's money to several billion dollars.  Success will depend on tapping space resources.  It will also depend on the development of spaceship building techniques and technology in microgravity.  Passage to Mars could use up a person's entire fortune.  Settlers could sell their house, cars and all other possessions to book passage to Mars along with the needed equipment for living and prospering on Mars.  Mars settlement would require some kind of organization and perhaps the sponsorship of wealthy individuals.  There have been suggestions to dismantle the ship once it reaches Mars.  In my view, expensive ships should be reused for as long as they will hold up and that could be many decades, even a century or more, in the rust and corrosion free environment of outer space.  Lightweight nuclear ships would not be able to land on Mars.  Settlers will descend from orbit in methane/LOX powered aerobraking capsules.  Early shipments to Mars will consist of machines for tapping Mars' resources to make methane and LOX and reusable single staged vehicles that can rocket up to orbit and return to the surface by aerobraking, parachutes and retro-rockets.

Lunar tourism is foreseeable but Mars tourism would only be for the richest people, at least until a large industrial presence exists in space with a thriving economy based largely on automation and teleoperation; and a large number of humans living, working and becoming prosperous in space.  Government backed efforts are not likely in free nations.  Colonization is a thing of the past.  Socialist experiments have no appeal.  Groups of settlers might form corporations and take their earnings based on their share.  Private homes and vehicles on the Moon and Mars seem reasonable.  What to do with the unemployed or homeless?  Find a way to make them earn their keep or ship them back to Earth!!!

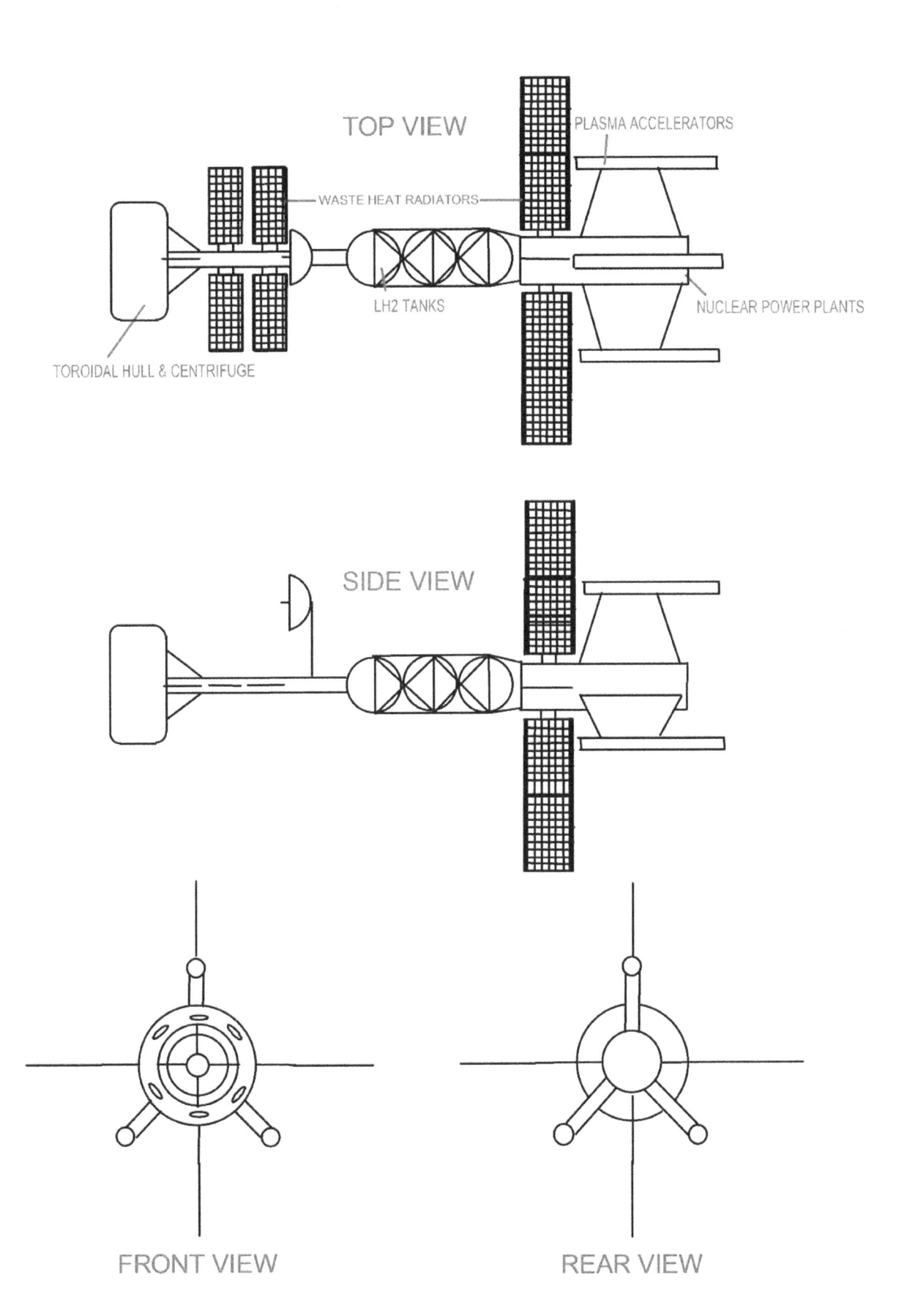

Fig. 15  Space liner for voyages to Mercury and the outer solar system.

# Space Corporations

Space will not be industrialized and settled by one giant over reaching government program funded by tax payers and the sale of US saving bonds. Numerous companies will be involved with each one or group of companies taking on a part of the job. At present, it looks like Blue Origin, SpaceX and United Launch Alliance will handle launching to Earth orbit, yet there might be some competition with other rocket launch companies. The first Moon mining base will probably be built by large energy interests that anticipate falling demand for fossil fuels in the future. All the hardware needed for the base including the solar electric tugs and disposable one-way lunar landers will be built by subcontractors. A vast amount of experience and talent in private industry will be tapped into. Corporate income tax breaks might be given to companies that invest in space industry projects. The Moon Mining Base company might be publicly traded too. There will probably be a lot of interest by millions of investors. There will also be naysayers who think climate change is a hoax and others who think ground based renewable energy is the future and look upon big business efforts to produce space energy with suspicion.

Once lunar materials are available, companies that build power plants like Bechtel, Shaw Group and Kiewit will construct GEO space stations and SSPSs. This will preserve them in a future time when large fossil fuel burning power plants are no longer built. These companies will either sell wholesale electricity to utility companies on Earth or sell the powersats to them. On Earth, power plants cost billions of dollars. Since loans have to be secured and stocks and bonds have to be sold to finance them, the long term cost also includes interest. Operating costs must also be considered. SSPSs will have no fuel costs. There won't be many moving parts to wear out. There is no rust, corrosion, storm damage, earthquakes, floods, wild fires, environmental lawsuits, socialist revolutions or terrorists in space. Solar panels may have to be replaced after a number of years due to degradation by heat and radiation in space; and microwave generator tubes may also have to be replaced after many years. The bulk of the powersat could last forever. Erosion by micrometeors just isn't that great. Labor costs for satellite operation over the years will also be low since most of the powersats' systems will be automated. Despite the high cost, SSPSs should be rather profitable.  Utility companies will also have to pay for rectennas and large scale power inversion devices.  The rectennas will be large screens made of aluminum wire that won't rust or block sunshine to farm land underneath.  Ground based solar will also have to include D.C. to A.C. power inversion so there's no advantage for ground based solar in that instance.

The excess materials from the Moon and oxygen from metal extraction might have to be dumped in the early stages of development.  Oxygen could be spewed into space if large tank farms for storing LOX don't exist.  Somebody is going to have to buy that stuff and make use of it for supplying propellant to spaceships and building LEO settlements like Kalpana Two.  Perhaps some smaller space hotels will be built first and the companies that do that will gradually grow until they can build Kalpana settlements.  Meanwhile, back on the Moon, the bootstrapping mining base will increase in size, replicate most of its machinery, boost output and maybe even complete railways.  It will also be possible to build hotels and resorts for tourists.  At first, modular towns consisting of 3D printed molten basalt habitat will be seen.  They will contain hotel rooms, restaurants, bars, pools, sauna, gambling rooms and farms to supply food and recycle air and waste.  Eventually, lava tubes will be sealed and pressurized with oxygen to build sub-selene communities for thousands of people and sports stadiums too.

The transportation industry in space will be a lot like the air travel industry today.  Airbus and Boeing are the two largest commercial airliner making companies.  Somebody will build a spaceship company and assemble sections of ships on the ground and pay Blue Origin, SpaceX, ULA or some upstart to rocket them to LEO.  The sections will be assembled in LEO.  The ships will be sold to spaceline companies that are perhaps part of or outgrowths of commercial airline companies now in existence.  We can imagine the names of some of these companies:

Night Sky

Silverado or Silver Moon

Dreamlines

Interlunar Inc.

Pegasus

PanSpace

Cosmic Travels

Alpha Corp.

Artemis

Etc.

Perhaps Virgin Galactic or Virgin Moon will be up there. A space station where sections of ships are assembled will be needed. Propellant depots and stations where travelers can transfer from vehicles that operate between Earth's surface and LEO to inter-lunar ships will also be necessary. Space stations will have to be located at EML1 or low lunar orbit (LLO). There will have to be lunar landers or "Moon Shuttles." The propellant depots might be the work of companies separate from the spacelines. Perhaps the SSPS builders will construct them. The space stations that serve as terminals in LEO and EML1 or LLO might be built by governments like municipal airports or train stations.

It seems like a lot of infrastructure but even the SpaceX Starship has to refuel upon reaching orbit thereby necessitating an orbital depot. Several 150 ton payloads of methane and LOX will have to be rocketed up to LEO at great cost and stored in the depot's tanks. The Starship can then carry 100 passengers to orbit, refuel and fly to lunar orbit. If the rocket with 380 second Isp methane/LOX burning engines must reach escape velocity, retro-rocket to the lunar surface and then lift off again and escape from lunar gravity, it will need a mass ratio of about 9:1. That will be impractical especially for a ship made of stainless steel. It will also have to hit the atmosphere at 25,000 m.p.h. and aerobrake. That would require massive heat shielding. Without infrastructure on the Moon to produce methane and LOX it won't be possible for the Starship to land and return to Earth. Besides, without hotels, where will the passengers stay? In the ship? That won't last long. Furthermore, there is not much carbon on the Moon so methane can't be produced. It looks like the Starship will only be able to fly around the Moon.

There might be ways to overcome this challenge. The Starship could land with enough methane for return flight and tank up on LOX made on the

Moon. Liquid oxygen composes 4/5 of the propellant combination so this could be advantageous. However, this requires LOX producing infrastructure on the Moon. The Starship could be a two stage vehicle, but then it leaves a stage behind on the Moon. Perhaps the spent stage could be scavenged for construction on the Moon.

There are over 46 million millionaires in the world today and about a thousand billionaires.[34] Once travel to orbit costs about $100,000 and flight to the Moon and back another $100,000 or so there will be plenty of people willing to pay for a trip to the Moon. Decades from today there will be even more rich people as the less developed countries industrialize and exploit their natural resources and labor. Wealthy people might forget about a second mansion or another yacht and shell out for space travel. While space construction projects might be done mostly by robots, spacelines will need humans to serve as spaceship pilots, flight attendants and spaceship service technicians. Orbital settlements will need farm workers, technicians and others. Plenty of jobs in space will be created as well as on the ground in rocket factories, even though automation will be used to keep cost and prices down. The spacelines could hold lotteries for people who want to win a free vacation on the Moon. Game shows might offer space and lunar vacations as part of prize packages. In another century there could be space travel prices low enough to be afforded by middle class citizens, but more importantly there could be opportunities for settlement on Mars and other worlds.

Many people dream of free space settlements large enough to house thousands, even millions, of people. We can only imagine building these with armies of AI machines. Materials would come from asteroids. The moonlets of Mars, Deimos and Phobos, might be mined to build settlements in Mars' orbit. It might be possible to spend a few months on Mars and a few months in an orbital settlement rotating fast enough to produce 1G without enduring major health deterioration.

Some imagine a swarm of space settlements orbiting the Sun housing a trillion or more people. This could be real in several centuries or several millennia. It depends on how fast the population grows and how fast giant settlements resembling steel balloons can be built by the AI machines. If the creation of a space faring civilization requires luxury travel for the elite to get started, who cares? I seriously doubt that an over reaching Russian government or People's Liberation Army program that provides good communist citizens with the reward of space tourism will ever work. High ranking communist officials would get to go and the only workers who get to go would just be used for propaganda purposes.

Life beyond money is also doubtful. Large space settlements might be built by developers who have claimed an asteroid and purchased a large number of AI machines. They could then build something like a Stanford Torus or Bernal Sphere and sell homes within to people who want to live in space. There would also have to be places to work in the settlements. Many people might move to a space settlement for a job while others go there just to retire. People might be somewhat different in the future. Biomedical science might extend lifespans, wipe out inherited disease and deformity and even boost intelligence. Longer lifespans alone will make people smarter by giving them more time to learn. Young people today are worried about getting a degree and starting a family before they are too old. If they could live say 300 years or more and remain fertile, they could find time to amass a great deal of knowledge and start a family later in life. They might afford several Ph.D.s thanks to on-line learning. With longer lifespans, people could save money, accumulate property to be sold later, invest in the stock market over a period of 100 or 200 years and become fabulously rich. In today's world you have about 40 years to work until you turn 65 and ten years later you are dead. That's not enough time to get rich through honesty, trading and hard work for most people. It's not enough time to explore the world and enjoy life either.

If lifespans are extended we can only hope that youth is extended too. Nobody will want to look like today's 80 years olds do for 200 years or more. A complete mastery of human biology and genetics would be required. Perfect control of human fertility would also be desired. People might reach maturity by age 18 but postpone child bearing until they are 150. If the time between generations remains approximately 25 years an uncontrollable population explosion would result and the resources of the solar system even would not be enough!

Besides education, there is experience. Someone who runs a business for a century or more could learn everything there is to know about directing a large corporation and then some. Imagine executives who know as much as all the head engineers and technicians. The only way businessmen are going to direct the activities of AI armies and a slew of human workers to build giant space settlements is to spend enough time working and learning to practically do all the jobs themselves.  The conquest of the solar system and then nearby stars would give multi-centenarians interesting things to do and not become bored with life and turn to decadence and debauchery even though there will be hedonists who choose that route.

In a solar system of a trillion or more people there would be billions of businesses.  A centrally planned economy would be absurd.  We might imagine a God-like computer that can do that, but I'd rather not.

Besides companies that mine the Moon, build SSPSs, build LEO settlements, produce, store and sell propellant at orbital depots, build spaceships, operate terminals and Moon Shuttles, build or operate hotels, resorts and railroads on the Moon, there will be companies that move passengers and cargo to Mars and other worlds, and do asteroid mining. The asteroid miners might not go into business until propellant is available in Earth orbit.  There might be companies that send out lots of small probes to sample and survey hundreds of NEOs and sell the information to asteroid miners.  Sections of actual AI asteroid mining ships might be manufactured on Earth because of their complexity and the rather crude manufacturing capacity in space and on the Moon.  They'd be rocketed to orbit and assembled in space.  The AI computers and software that navigates the ships, operates the mining and processing machines, and steers the ships back to Earth-Moon space might be the most expensive components of the mining ships.  All sorts of computers will be needed in space for navigation, telecommunication, environmental systems control, construction robot direction and many other purposes.  Special software will be needed for them all, so the IT companies will make lots of money selling commercial software for a pretty penny.

The most common type of asteroid is the C-type asteroid and many of these contain water and organics.  As industry grows in space the demand for these will skyrocket.  There will also be a market for platinum, iridium, other platinum group metals, gold and silver.  The problem is that if there is too much competition and too many asteroid miners with a lust for precious metals, the supply could grow to high and prices could plummet, thereby taking all the profits out of mining metallic asteroids.  Perhaps the high powered lasers needed for metallic asteroid mining will be so expensive that only the richest individuals and corporations could afford them and this will naturally keep over mining for precious metals in check.

The ships to Mars will also be made on Earth and assembled in space. All these interlunar, asteroid and Mars vessels will require a sort of orbital ship-yard or assembly space station, maybe several of them. Interlunar ships with centrifuges might be modified to run nuclear electric drives so they can travel rapidly to Mars. Cargo ships could run nuclear electric drives or they could use chemical propulsion that can send them on Hohmann trajectories to Mars in 260 days. The cargo ships would be robotic so rapid transit won't be important. The assumption here is that chemical rockets and propellant will be cheaper than nuclear electric systems and hydrogen. Simple pressure fed silane/metal powder and LOX fueled rockets with moderate thrusts seem as if they would not be nearly as expensive as vapor core reactors with magnetohydrodynamic and supercritical $CO_2$ turbines for electric power generation and superconducting VASIMR rocket engines.

Martians will need all kinds of cargo. They will need plastic greenhouse domes made of Kevlar, excavators, compressors to pump down atmospheric gases, electrolysis units, fuel cells, batteries, solar panels galore, gas liquifying refrigeration machines, storage tanks, brick making machines, metal smelters, small 3D printers and large gantries for making habitat out of melted extruded regolith, heating systems, provisions of food, water, oxygen, medicines and clothing, spacesuits, seeds for crops and eggs for various fish, chickens and insects, ground vehicles, inflatable habitat, computers, radio communication equipment, hand tools and machine tools, robots and more. This will all be needed just to get started. Farming will have to be successful before provisions run out. Habitat will have to be made of extruded regolith and bricks held in compression by tons of regolith. Robots and a small crew will have to go first and prepare a base for larger numbers of settlers later on. Many companies will make these things on the Moon, in space and rocket them up from Earth too, for a price, of course. Many merchants on the old western frontier made more money selling supplies to prospectors than the prospectors ever did digging for gold. There will be plenty of merchants in the future willing to sell things to the Martian settlers.

Transportation to Mars won't be cheap. A ship that costs over a billion dollars that could haul thousands of people to and from the Moon every year would only be able to carry some people to Mars in 39 days then return to Earth in a similar amount of time. In that much time a ship could haul 2000 people to and from the Moon for at least $100,000 each and earn $200 million. Who would build a ship to Mars when more money can be made from the investment hauling tourists to the Moon unless the passionate would-be Martians were willing to pay as much if not more?? It wouldn't make sense to dismantle a billion dollar ship even if it could land on Mars. Could the Mars settlers themselves pay for the ship and use it over and over again to haul future settlers to Mars? Could they buy an old second hand lunar tourist ship and outfit it with nuclear electric propulsion? They'd have to modify and improve the life support system to so that passengers could be sustained for 39 days in a row instead of just a few days. Where there's a will there is a way and souped up used spaceships and modified robotic asteroid mining ships for cargo might make the dreams of the Martians affordable someday, but not anytime soon.

The red planet settlers might pool all their money and share in the wealth of the settlement. This might involve forming a corporation and sharing of the wealth based on percentage ownership of shares. A wealthy benefactor who wants to pioneer on Mars might toss in a billion dollars and just ask for a slightly larger percentage so he can live in the biggest house on Mars or afford the most kids. If Arabs did this they'd want a big enough percentage to afford more wives as well. Meanwhile the communists, if they are still around in the future, will be selecting their best citizens to be comrades on Mars and prove that socialism is superior to capitalism. The Russian or Chinese government would pay for everything, presuming some capitalist will sell them the spaceship, so that the workers can be proud of their nation. Let's hope for a better, freer future than that.

# Space Ports Etc.

A small stout spacecraft for moving passengers to ELEO doesn't have to be very large because travelers won't be on board very long. If this is true, why do spacecraft going to the ISS have to take hours, even days, to rendezvous with the space station? That's because the ISS is in an orbit highly inclined to the equator. Rockets launching from Kennedy Space Center or Russia also enter highly inclined orbits. If the launch bases are located on the equator and the space stations are put in Equatorial Low Earth Orbit, the orbital mechanics are much simpler and launch windows from the ground to the stations occur every 100 minutes. It makes sense to put the launch bases on the equator, on the shores of the Indian or Atlantic Oceans, and put the space stations in ELEO. This would place the launch bases in Brazil and Somalia. Borneo is also on the equator with the Pacific to the east. Cargo launches could be done from bases at higher latitudes, but scheduling tourist flights to the space stations would be simplified if bases are located on the equator. If there is a medical emergency, patients could be returned to Earth more rapidly.

Space tourists will have to fly to distant parts of the world to get on board a rocket. Airports and hotels would have to be located nearby. Rocket manufacturing and servicing facilities would be needed. The US Department of Defense might not like it, but military payloads can always be launched from KSC or Vandenburg. The aerospace machinists' union might not like it either, but people looking for work in these distant nations would be pleased.

It only takes about ten minutes to reach LEO. Passengers will spend no more than an hour in the orbital vehicle. Once they reach a space station they will either stay there and vacation, or board a ship to the Moon. The stations would consist of finished modules and inflatables launched up from Earth and sections made from external tanks. Eventually, there will be large Kalpana stations. In ELEO at an altitude of about 300 miles (500 km) there is a slight amount of air friction, so there will be electric drive units on the stations or electrodynamic tethers to provide a small constant thrust to overcome that friction and prevent orbital decay.

Travelers to the Moon will transfer to a ship docked at the station. The inter-lunar ships will be very lightly built and they will never land. They will only travel from Earth orbit to lunar orbit and back to Earth orbit. The mass of the ships must be kept down for fuel efficiency. About 200 passengers with baggage would only weigh about 20 metric tons while the ship and its propellant will weight several hundred tons. The ships must be large enough to accommodate all these people for 30 hours to 80 hours. Several hundred or more tons of propellant will be needed for these ships. A propellant depot will orbit nearby. It could be built of machinery launched from Earth and external tanks. A cluster of tanks would be located on the end of a boom. The depot would rotate slowly to produce a small fraction of one Gee, perhaps just milli-Gees, to settle propellant in the tanks. Pumps would transfer the liquid oxygen and silane via flexible metal pipes to an inter-lunar ship docked on the other end of the boom. Refrigeration machinery would reliquefy boil-off from the depot and ship tanks. Solar shields would be needed to prevent heating of the supercold fluids. Shields to prevent Earthshine from heating the tanks would also be needed. Solar panels and batteries would supply power.

Propellant will come in the form of liquid oxygen, metals and silicon tanked down from SSPS building stations and hydrogen piggybacked with payloads from Earth. The orbital depots will also need silane production systems. Designing and building this equipment will keep chemical engineers quite busy. The process is fairly simple. Magnesium and silicon powders are reacted at high temperature in electric furnaces to produce magnesium silicide. This is then reacted with hydrogen chloride to form gaseous silane and solid magnesium chloride. The hydrogen chloride is made by combining hydrogen with chlorine gas. The reaction is very exothermic. The magnesium chloride can be put through electrolysis to recover and recycle magnesium and chlorine. The heats of reaction and the corrosive nature of the reagents will give the chemical engineers headaches, but that's what they get paid for. Mixing the liquefied silane with metal powders will be one more not insurmountable challenge.

The propellant making stations/depots will be complex and costly and that cost will be added into the price of a ticket to the Moon or passage to Mars. Hopefully, the cost of this infrastructure will be worth it to the businessmen who try to make space tourism affordable and profitable. There will also have to be space stations and depots in lunar orbit where passengers transfer to landers or Moon Shuttles and ride down to the surface. The ships would refuel for return flight. Loads of material could be hauled by mass catchers from L2 where it could be processed at the lunar orbital stations/depots. Hydrogen from polar ices could be rocketed up from polar mining bases. Much of the lunar stations could be launched from Earth. Materials from the Moon could also be used to bootstrap them up. There might even be SSPSs in lunar orbit or at L1 to beam power down to the Moon.

**Moon Lander**

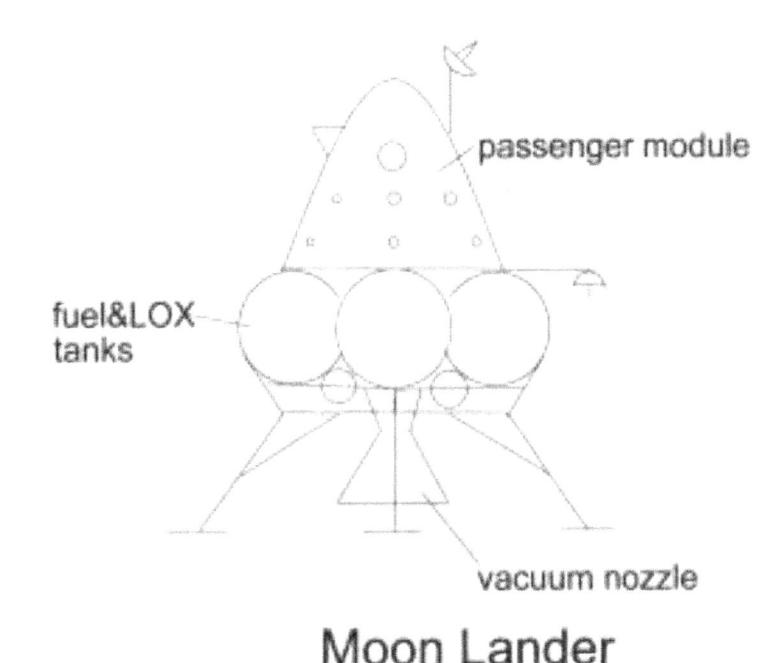

Fig. 16 A Moon Lander or Moon Shuttle

The Moon Shuttles could use silane, powdered metals and LOX or they could use a monopropellant made of metal powder and LOX. They will have to be stout enough to endure lunar gravity and somewhat more during ascent and descent. Moon Shuttles would be built on the Moon with electronic parts imported from Earth. They could provide rapid suborbital travel across the Moon as well as hauling people and more delicate cargos between the lunar surface and lunar orbit.

The surface bases will have large areas of microwave or electron beam sintered regolith to provide a solid landing surface. This will prevent sand blasting of anything nearby when Shuttles take off or land.  There could also be a barrier of bulldozed regolith surrounding the pads.  These bases will probably be located in the smooth maria instead of the rugged highlands.  There are bound to be many places that are too rough for spacecraft to land in.  Surface vehicles and railways will be needed to access areas strewn with boulders and uneven terrain.  It would also make sense to locate the launch and landing pads several kilometers away from the Moon bases so that an off-course rocket doesn't crash into habitat and cause disastrous depressurization.

Spaceships, Moon Shuttles and all the infrastructure associated with them will certainly require an enormous investment.  If bootstrapping a mining base with mass driver, mass catchers, GEO construction stations and SSPS building come first it should be possible to make profits in a reasonable amount of time.  Perhaps twenty year bonds will be sold.  Much of the same infrastructure used for powersat building could also be applied to lunar tourism infrastructure.

Lightweight ships to Mars won't be built to land either.  Small spacecraft or "Mars Shuttles" will haul passengers and cargo down from Low Martian Orbit (LMO) and back up again.  These vehicles will probably be propelled by methane/LOX burning rockets made on Mars where there is plenty of carbon in the atmosphere and polar ice caps.  The Mars Shuttles will use aerobraking and parachutes to land without so much propellant consumption.  The moonlets, Deimos and Phobos, offer resources.  It is believed they are chondritic but not carbonaceous.  Even so, they probably consist of silicates that could be a source of oxygen.  Simple magma electrolysis could extract that oxygen.  Shuttles could reach orbit with just enough methane for retro-rockets and take on LOX in orbit.  The moonlets might contain iron, magnesium and silicon dioxide (glass) for construction in space and they might be used as bases for orbital activities.

It should be possible to fire tanks of liquid methane up to Mars orbit with mass drivers located atop equatorial volcanoes well above the denser layers of atmosphere. With oxygen in Mars' orbit it will be possible to load up booster rockets with propellant and move nuclear electric ships up to escape velocity in short notice.  The NEP ships will thrust for weeks and reach very high velocities for quick return to Earth or travel to other destinations in the solar system.  The Main Belt asteroids, the moons of

Jupiter and the moons of Saturn are all more accessible from Mars. Gravity assist from Jupiter fly-by might be used to speed up ships headed for the outer solar system.

The Saturn system is particularly interesting. Titan, the only moon to have an atmosphere, could be explored with airplanes. Spacecraft could aerobrake and parachute or glide down to Titan's surface and use jet-atomic propulsion to climb back up again. Methane from Titan's atmosphere and lakes could be used as reaction mass in nuclear thermal rockets that travel throughout the moons of the Saturn system. Hydrocarbons from Titan and ices of various Saturnian moons could supply space settlements in orbit around Saturn and in the outer solar system with light elements including nitrogen.

| moon | dist. | diameter | T.O.F. | synodic period | dV |
|---|---|---|---|---|---|
| Janus | 151,472 km | 95 km | 3.8 days | 0.725 days | 6.7 kps |
| Mimas | 185,600 | 390 | 3.94 days | 1 day | 6 kps |
| Enceladus | 238,000 | 500 | 4.15 days | 1.5 | 5.2 |
| Tethys | 294,660 | 1050 | 4.375 days | 2.14 | 4.45 |
| Telesto | 294,660 | 15 | 4.375 days | 2.14 | 4.45 |
| Calypso | 294,660 | 15 | 4.375 days | 2.14 | 4.45 |
| Dione | 377,400 | 1120 | 4.7 days | 3.31 | 3.6 |
| Helene | 377,400 | 32 | 4.7 days | 3.31 | 3.6 |
| Rhea | 527,200 | 1530 | 5.34 days | 6.3 | 2.5 |
| Titan | 1.2 million | 5150 | --- | | |
| Hyperion | 1.5 million | 250 | 9.9 days | 63 | 0.5 |
| Iapetus | 3.6 million | 1460 | 22.7 days | 19.88 | 2 |
| Phoebe | 13 million | 220 | 113 days | 16.37 | 2.9 |

**FROM TITAN TO OTHER MOONS OF SATURN**

Fig. 17 Minimum energy trajectories from Titan to other moons of Saturn.

# Farming on the Moon

It has been argued that lunar agriculture is not feasible primarily because of the power demand for crop illumination. Greenhouses will need thick glass roofs, crops will be killed by solar flare radiations and overheating of the greenhouses during the two week-long lunar day will occur. None of these arguments are valid.

Thick glass roofs will not collapse in the low gravity of the Moon. Greenhouses will be exposed temperature extremes that will cause expansion and contraction of materials that could lead to cracking and other structural failures. Micrometeoroid damage is another hindrance. The fact is, greenhouses will not be used on the Moon. Freshnel lens collectors or "heliostats" and light pipes will transmit UV filtered sunlight to farm chambers in modular habitat.  The early modules will be inflatables from Earth covered with several meters of regolith for thermal insulation and radiation shielding as well as micrometeor protection.  Later modules made on the Moon will consist of metal plates welded together or they will be 3D printed with extruded molten basaltic regolith.

In the more distant future, light will be piped into sub-selenar lava tubes which might be hundreds of meters in diameter and many miles long. Farms and buildings will be established within the tubes.  Areas inhabited by humans in these lava tube cities will also be lit by light pipes and skylights. Underground farms will enjoy a constant temperature of a few degrees below zero Fahrenheit in the surrounding rock rather than 250°F. days and minus 250 °F nights. It will be easy to warm the chambers up to 72 °F. with waste heat from nuclear reactors or molten salts stored up with solar thermal systems during dayspan. Radiation from galactic cosmic rays and even the strongest solar flares will be no problem deep beneath the Moon. Micrometeoroid punctures will be unheard of. Overheating or "supergreenhousing" will not occur.

Illumination during the two-week long lunar night could be produced by microwave sulfur lamps with flexible fiber optic light-pipes that direct the light to the places where it is needed most. Light will not simply be scattered all over the place to be absorbed by the stone walls. Sulfur lamps provide light in the visible range with very little infrared or ultraviolet. These revolutionary light sources can produce 95 lumens per watt.[35] Incandescents yield only 20 lumens per watt and fluorescents give 50 lumens per watt. Sulfur lamps don't even have electrodes to burn out!

Plants don't use the entire light spectrum. Their chlorophyll absorbs red and blue light mostly. Red and blue LEDs would be the most efficient way to illuminate crops during lunar nightspan.

In the past, illuminance recommendations were not as high as today's. In 1925, A Text-Book of Physics suggested that night time street lighting required less than one lumen per square foot. The average living room only a few lumens per square foot. Offices and classrooms needed 5 to 10 lumens per square foot. Workplaces where fine handicrafts, engraving, sewing or drafting were being done needed 10-20 lumens per square foot.[36]

Today, we find values of 75 foot-candles ( one ft. candle= one lumen/ sq. ft. or 10 lux) for reading and office work, 50 ft. c. for machine operation and 50-300 ft. c for bench work.[37] The noon-day Sun gives off 10,000 lumens per square foot at Earth's surface!

Plants need more light than humans and animals do, but not this much. Many plants only need 200 lumens per square foot for good growth! The small tropical Chinese Evergreen plant, Aglaonema modestum, only needs 100 lumens per square foot ( same thing as 100 foot candles) and can get by on as little as 10 lumens per sq. foot.[38] The Bamboo Plant, Chamaedorea erumpens, requires just 100 to 150 foot candles. The coffee plant, coffee arabica, a necessity for us groggy old lunar prospectors and rich travelers, needs 150 to 1,000 lumens per square foot.[39] Tomatoes, sweet peas and everbearing strawberries need 1500-2000 foot candles and cucumbers require 4000 foot candles.[40] If these plants receive 1500-4000 lumens per square foot from free sunlight during the lunar day and just 1000 foot candles for 16 hours out of every 24 hour period from sulfur lamps during the lunar night they might do just fine. A thousand foot candles is like a cloudy day.

Although the Sun might drench the Earth with the energy of 4 MW per acre, 1000 MW per square kilometer, and 2500 MW per square mile, only a tenth of this is needed for light hungry plants like the coffee plant. A one acre garden plot in a lunar lava tube illuminated by sulfur lamps will need 43,560,000 lumens to deliver 1000 lumens per square foot. Only 460 kilowatts will be necessary for one acre if sulfur lamps rated at 95 lumens per watt are used. To illuminate a square mile of lunar gardens, 290 megawatts is needed. This is not impractical given the intense, constant solar energy that's never obscured by clouds available by dayspan on the Moon that can be harvested with silicon solar panels or polished magnesium solar thermal collectors and stored in the form of hydrogen and oxygen that can energize fuel cells for electricity by night. Nuclear reactors can also be used on the Moon with impunity. There is no air, no groundwater, no wildlife and no ecosystem on the Moon that could be harmed by a meltdown or nuclear waste dump. Nuclear fuel could be reprocessed and breeder reactors could be used to tap the energy of plutonium from U238 and U233 from thorium 232. Massive containment buildings won't even be necessary. Terrorists will never make it to the Moon and if they do they will never make it back to Earth.

Although we can generate the electricity needed to furnish the crops with light, there are many other strategies to make lunar farming successful. Mushrooms can be raised in the dark. Three pounds of edible fungi per square foot of garden space can be harvested every fifteen weeks.[41] Algae like Spirulina can be cultivated during lunar dayspan. Since blue-green algae can double its mass four times a day (24 hour period), in five days 100 grams of algae could reach a mass of 100 metric tons if it has enough water tank volume, minerals and carbon dioxide. It is therefore possible to grow enough algae while free sunlight is available during the lunar day to feed livestock throughout the month. Fish can eat algae. Goats and pigs will eat anything.

Algae is actually very nutritious; high in protein, minerals and vitamins. Chickens might eat pellets of algae. Mixed with dirt, algae could supply carbohydrate for mushrooms. Mushrooms could feed the animals too. Moon dirt could be mixed with algae and mushrooms, allowed to rot and form a rich compost, and earthworms could be farmed in the rich dirt. Chickens and fish will eat chopped worms. Livestock won't need more than a few lumens per square foot to see. Fungi and worms won't need any light and algae only needs to grow by day. Clearly, a lunar diet rich in mushrooms, fish, chicken, eggs, pork, goat meat, goat's milk, cheese, butter and cream can be produced without artificially illuminated crops at all!

Eggs and liver are rich in vitamin A, so no one will die due to a lack of carrots. Meat has plenty of B-complex. Milk contains vitamin D or people can just sunbathe for 10-20 minutes a day. Some vitamins C and E are still desired, and so is some fiber. Some wheat for whole wheat bread and dough, tomatoes, potatoes, lettuce, grapes, strawberries, cucumbers and pumpkins can be grown with sulfur lamp or red and blue LED illumination by night.

A diet heavy on meat, fish and dairy products consumed during a two-week vacation on the Moon will not irreparably damage anybody's coronary arteries. Hotel workers, miners and scientists spending a couple of years on the Moon won't die of heart disease either if they stay fit. Fish, chicken and lean goat chops might be preferable to lots of eggs, heavy cream and bacon for the health conscious Lunans.

The Moon will never support billions of people as Earth does or Mars could after centuries of terraforming, but it doesn't have to. Millions of miners, scientists, workers and tourists who are the life blood of the Moon can be supported by the underground farms in lava tubes and man-made tunnels that will someday be planted in the Moon. Eventually, craters will be domed over with giant bi-layer silicone bubbles with seven meter thick water shields for radiation protection. Fusion powerplants "burning" helium 3 will supply electricity for the sulfur lamps and the resources of near Earth asteroids will be utilized. Subway tunnels will interconnect the domed craters. The Moon will become a fantastic playground and a jewel for all citizens of Earth, like the Great American West today.

# Preventing Pandemics in Space

In the closed environment of spaceships and space habitat the potential for widespread infection amongst space workers, tourists and others is very high. Steps must be taken to prevent epidemics. All workers could be quarantined for 14 days or longer before launch. They would have to be tested for everything from TB to Covid-19 and whatever new mutant viruses and bacteria emerge in the future. They would need vaccination against a large number of contagious diseases from multiple influenza strains and pneumonia to Covid-19 and tuberculosis. Old disease threats as well as new ones would endanger humans in space.

Tourists might resist going into quarantine for 2 weeks or longer before launch. They will have to have a full battery of shots and tests or they will be kept off spacecraft; forcibly if necessary. No good immunization and test records—no ticket, no passport and no exceptions. If a tourist comes down with a cough, cold, fever or digestive illness a few days before launch they will have to have their trip postponed and their money refunded if need be. We cannot take chances with a minor illness turning out to be potentially deadly. In all probability, most people will be reasonable about these precautions. Service crews will deep clean and sterilize spaceships interiors with dry steam before and after every flight. Things might get wet but the hot moisture left behind by steam evaporates quickly. Powerful dehumidifiers will condense the water from the air. No noxious chemicals like chlorine bleach that would poison the air would be used to clean interiors. Hydrogen peroxide which decomposes to water and oxygen could be used to disinfect interiors. Ultraviolet light might also be used to sterilize interiors as is done in some hospital rooms today.

At first it would seem that workers at early Moon bases would be housed in "tuna can" habitat where space is at a premium like in a submarine. They would sleep in bunks, possibly shared bunks for workers on different shifts, and have a common bathroom with showers, hand basins and toilets. This would be a real breeding ground for contagious disease. Even with quarantine before liftoff, shots and testing, an evil microbe might get through and that just cannot be allowed.

If private cabins with private baths can be had with air ventilation systems that create negative pressure in cabins, the cabins could also serve as isolation units if someone gets sick. Air vented out of the cabins would be cleaned with HEPA (High Efficiency Particulate Air filters) filters that can remove at least 99.97% of dust, pollen, mold, bacteria, and any airborne particles with a size of 0.3 microns (μm) or more.[42] Wastewater would be filtered, distilled and sterilized with heat, pressure, UV light and perhaps small amounts of ozone and chlorine.

It would seem that shipping this extra mass of habitat instead of "tuna cans" would cost a lot of money. Would AI robots be smart enough to build roomy habitations meeting all these requirements with local resources like basalt bricks, metal plates, glass, etc.? Or would it be possible to deploy spacious inflatables? Inflatables have much better volume to weight ratios than hard modules and they are easy to deploy—just pump them up with air. Plumbing and air ventilation systems would consist of flexible plastic hoses that are folded up with the rest of the inflatable when it is packed into the nose of a rocket. Even with a pressure of say 10 psi (3 psi oxygen) they would have enough internal pressure to support a layer of regolith several meters thick in low lunar or Mars gravity for micrometeoroid, thermal and radiation protection.

Submarines stay down for about 70 to 90 days at a time. That's about as much as the crew can stand psychologically. Given the high cost of transport into space, to the Moon or Mars, work crews may have to spend much more time off Earth. If they can have their own cabins, private baths and roomier common areas for eating, cooking, exercising, recreating, etc. they can endure life in space more easily. Shared cooking and eating areas might seem like disease breeding grounds but extra care would have to be taken to keep everything clean. Dishes could be washed in dish washing machines with extra hot water and soap. Soap kills viruses. It might even be possible to make dish washing machines with UV sterilizing lamps within. Kitchen areas, tables, chairs and such would be wiped down with soap and water, hydrogen peroxide solution and maybe some pure ethanol. They might be steam cleaned from time to time. UV light sterilizers could also be used at night in the kitchens and dining rooms when everyone is asleep, depending on how shifts are staggered.

Air would be sucked out of cabins with powerful fans to create negative pressure. Air would enter the cabins thru vents with micro-filters in the polymer sealed Kevlar fabric walls of the inflatable habitat. Any bugs that get in won't get out. Dirty clothes and bed sheets would be washed in machines with soapy water and UV sterilized. This is preferrable to the use of strong chlorine bleach because that could contaminate the recycled water and kill yeast and bacteria in the bioreactors that degrade wastes. Air from the cabins could go thru a bed of red hot gravel or metal wool, washable reusable micro-filters and UV lights. This would prevent the spread of respiratory diseases.

There are commercially available air filtration systems that use six stages of purification: **1.** Negative Ion Generation Technology **2.** Adjustable Ozone Generation up to 500 mg/hr. **3.** Activated Carbon Deodorization **4.** UV Light **5.** Photocatalytic hydrogen peroxide generation **6.** Washable HEPA filter to eliminate small particles and dust. [43] This technology might be "just what the doctor ordered."

Personal protective equipment would also be needed, especially for doctors and nurses in sick bay. At first it could be shipped to the Moon or Mars. When material production and factories come on line PPE could be made with 3D printers. PPE could be washed and reused and be treated with UV light if it is made of the right kind of plastic and it could even be ripped up, ground down, melted, drawn into strings and new sheets of plastic could be printed. Paper micro-filters for masks might be made from bamboo. Sick bays would need the latest equipment including oxygen and respirators. It will be wise to plan and prepare for any medical emergency. Of course, sick bay will be maintained at negative pressure.

Eventually, habitat will be made from local resources like basalt, glass and metal. Furniture made of basalt and metal with woven basalt fiber cushions stuffed with basalt fiber could easily stand up the heat of dry steam and exposure to UV light. Many plastics would be degraded by those.

# Off-Earth Mining: Speculations

It is thought that the Moon lacks ore bodies. Many ores and mineral veins on Earth are formed by hydrothermal or sedimentary processes. Alas! The Moon never had any water, but Mars did. Some speculate that there may be gold veins and other deposits on Mars. The Moon's surface has been bombarded by meteorites for eons. It has been pulverized into a fine powder and a lot of mixing has occurred. Subsequently, it is thought that the Moon is covered by a fairly homogenous regolith and dense concentrations of minerals are unlikely. However, what about the deeper rock underneath the regolith?

There are volcanic domes on the Moon, pyroclastic glass deposits from volcanic vents, and lava outflow channels. The nearside of the Moon has not been volcanically active for 3 or 4 billion years. On Earth there are magmatic nickel-copper-iron-PGE (platinum group element) deposits and volcanic hosted massive sulfide (VHMS) copper, lead and zinc deposits.[44] Copper and zinc are needed for high strength 2000 and 7000 series aluminum alloys. Zinc improves the ductility and machinability of magnesium. Could there be deposits of these valuable elements in the Marius Hills and Aristarchus Plateau where volcanism is known to have occurred?

There is no plate tectonic activity on the Moon. Volcanoes could only form over hot-spots where molten material rose up through the mantle but cooled off billions of years ago. The only other possibility for heating would be the decay of radioactive minerals. Apollo 15 detected an increase in alpha particles when it flew over Aristarchus in 1971. Radon 222, a decay product of uranium and thorium, has been detected over Aristarchus by Lunar Prospector.[45] Could there be an anomalous concentration of uranium and thorium beneath the Aristarchus Plateau? Or could radioactive minerals have been brought upwards with the hot spot molten material? There is really no way to know without more investigation including deep drilling.

If we discover deposits of uranium and thorium richer than KREEP, this will benefit space exploration and settlement. Launching nuclear fuels from

Earth is preferably avoided; however, nuclear power is needed for spaceships with electric propulsion like VASIMR for travel to Mars and beyond. Nuclear power could also be used for polar crater ice mining machines. The only alternatives would be long tethers or beamed power.

Another interesting thing about Aristarchus is that it is one of the few places on the Moon where granite exists. Granite forms on Earth near subduction zones when magma rises up and gradually cools and forms a series of minerals. The granite forms underground and cools slowly with the formation of coarse grains. If this material reaches the surface it is called rhyolite, a fine grained mineral. Erosion and geological upheaval can later unearth the granite.

A RICHER SOURCE OF
NUCLEAR FUEL THAN KREEP?

ARISTARCHUS CRATER

granitic magma

andesitic magma

basaltic magma

MINE SHAFT

Anomalous concentration
of radioactive minerals
PAYDIRT!

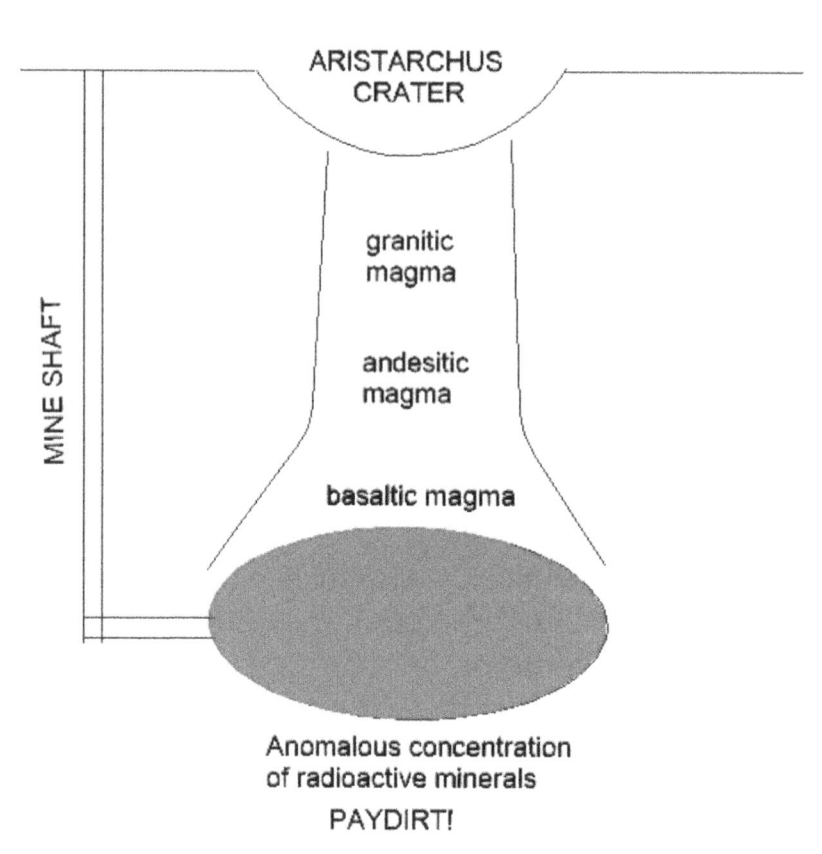

Fig. 18 Pure speculation, but what if???

We can engage in more speculation than this. The mineral chromite, $FeCr_2O_4$, a rich source of chromium, is dense and sinks in magmas. In this way, layers of chromite formed in the Bushveld Igneous complex in South Africa. The seas of the Moon, the mare, were once molten lava filled plains. Could chromite, which is found in small amounts in the mare, have sunk and formed mineable layers beneath the surface? Chromium is needed for stainless steel and synthetic rubies. This makes us wonder if large high energy laser crystals could be grown in the low gravity and free vacuum of the Moon? NASA's Moon Mineralogy Mapper (M3) on India's Chandrayaan-1 detected large amounts of chromite in the Sinus Aestuum plains right in the center of the nearside. The chromite spinel is scattered over tens of thousands of square kilometers and it is believed to be of volcanic origin and exhumed from deep within the Moon.[46]

Mining often requires explosives to blast into hard rock. There is no oil or ammonium nitrate on the Moon and not much carbon and nitrogen to make nitroglycerin. Mars does have hydrogen, carbon and nitrogen with which to make chemical explosives. Despite the lack of elements for conventional explosives on the Moon, it will be possible to use oxyliquits.[47] These are explosives made of containers filled with LOX and aluminum powder that can be ignited with an electric spark. Mixtures of LOX and aluminum have been shown to make monopropellants, but magnesium in LOX is shock and vibration sensitive and detonates. Perhaps magnesium could be used as an explosive if it is handled gently.

Explosives could be used to blast boulders out of the way, make road cuts by blasting through rock, tunnels through crater rims, cuts and fills in valleys for roads, create entry ways into intact lava tubes, dig mine shafts and perhaps other jobs. If there are ore deposits in lunar rock, it will be possible to mine them. More exploration is necessary.

# Making it in Space

David J. Gingery wrote and published a 7 book series called "Build Your Own Metal Working Shop From Scrap." He starts with a charcoal foundry and sand molding to make parts for machine tools. The castings are finished by hand at first, but once some tools are built the machines are used to make their own parts. He describes making a metal lathe, a metal shaper, milling machine, drill press, dividing head and sheet metal brake.

On the Moon, high orbit or on Mars we won't find scrap metal lying around, with the exception of external tanks in orbit, or use charcoal foundries, but we can use devices similar to mass spectrometers or calutrons, like Dr. P. Schubert's Supersonic Dust Roaster and All Isotope Separator to extract metals from regolith. The metals can be melted in electric furnaces and poured into sand molds bound with sodium silicate since water and clay won't work in the vacuum, and resins would be in short supply. The sand molds could be made early on in pressurized inflatables by human workers.

Sand molds require patterns. These are often made of wood shaped by skilled artisans. Off-Earth, we will use plastic and 3D printers. Some plastic patterns will be imported, but many will be made on site. Old patterns will be ground up, melted and extruded into new filament for the printers.[48] It should be possible to make any pattern desired and many patterns will be reused to make numerous sand molds.

We will bring metal lathes, drill presses, milling machines, etc. with us into space. We will also have various 3D printers and CNC machines. The task of replicating all the machinery will be much easier with these machines than by doing everything by hand. We won't stop at replicating machine shop tools. Large rolling mills including the big heavy rolls and frames will be made by sand casting steel and grinding things to exact tolerances with CNC machines. Smaller parts will be made by 3D printing and machining parts in the shop. Some imports, like electronic controls, could also be used. The rolling mills will make it possible to produce metal plate, sheet, foil perhaps, rails, beams, bars, rods and pipes, mostly from aluminum and magnesium alloys.

Sheet metal will be very important in high orbit. SSPS frames will be made of beams made of sheet aluminum and reflectors will be made of magnesium sheet coated with a thin layer of aluminum perhaps. Sheet metal can be used for all kinds of things from tool boxes and shelves to solar shields. Metal (iron or steel) rails will be needed on the Moon and Mars in huge quantities for railroads. Shipping rolling mills which can weigh 20 to 40 tons to the Moon, high orbit or Mars, in the quantities needed would be far too expensive. A series of half a dozen or more of these machines would be needed to roll plates and sheets. That would weigh hundreds of tons. Better to make them on site.

Bootstrapping is all about using the tools we bring with us to make the tools to make the tools to make the needed items. The SDR-AIS is used to make the metal and that is used along with sand molds and tools to make the mills that make the sheets that are used by the beam builders to make the beams that are pieced together to make the SSPS. On the Moon we will want sheet metal or foil reflectors to concentrate solar energy on imported multi-junction solar panels to increase electrical power output. Metal plate, tubes, beams and other parts could be used to make ground vehicles, excavators and spacecraft like Moon Shuttles and Mars Shuttles. Gantries, basically huge 3D printers, for making habitat from melted and extruded regolith will be imported at first. Later, we will make them on the Moon to expand habitat production and build large research and industrial bases, hotels and resorts. The same could be done on Mars.

Skilled humans casting and machining parts and assembling them will be indispensable. They will build robots for space construction with local materials combined with sophisticated electronics from Earth. Eventually, there may be factories for making electronics including computer chips in space. Artificial intelligence will supplement human workers.

Relativity Space uses a huge 3D printer with AI driven controls called Stargate to actually print rockets faster than can be done by traditional manufacturing.[49] Devices like Stargate could be taken to the Moon, Mars and high orbit where they could print rockets, vehicles, spacecraft and other machines. The digital controls would be sent up from Earth while many other parts of the Stargate could be made off-Earth by skilled workers. Eventually, the Stargate might print all the parts needed to replicate itself. Assembly would be done by humans and robots working together. In the more distant future, advanced AI could replace humans entirely.  Manufacturing and construction off-Earth could be 100% automated.

With advanced AI it will become possible for machines to be sent to other worlds where they will build everything needed for human habitation in advance of arrival by explorers and settlers.  In this way we might expand into the solar system and even reach the stars.

Presently, AI hasn't replaced humans and probably won't for some time.  It's not possible to see the future, but the incredibly rapid progress of digital technology over the past few decades convinces us that AI will meet or exceed expectations.  This raises the question, is it wiser to wait until AI can do all the work we want done in space, or should we embark on a space industrialization and settlement program requiring humans now?  To answer with more questions, how long do we want to wait to make money in space and how long do we want to postpone our energy future with climate change worsening every day?  Are the risks to human life in space worth taking?

# Hope for Mars?

According to Carl Sagan, in his book *Cosmos,* the polar caps of Mars could be dusted with black material so that they absorb more solar radiation, warm up, melt and vaporize. This would thicken the atmosphere of Mars, cause it to trap more heat, warm up the planet and even cause the melting of sub-surface permafrost. About 1,200 Saturn V payloads would be needed to do this.[50]

Perhaps it would be more efficient to establish bases on Mars. Dark basalt could be mined from volcanic regions and launched by sub-orbital rockets or mass drivers to dust the polar caps and induce warming. Other methods of terraforming Mars have been proposed. Detonating thermonuclear bombs over the polar caps is out of the question. Radioactive fallout would poison the red planet. Small asteroids could be crashed into the polar caps and the energy released upon impact could melt the ice. A large asteroid might result in explosive blow-off and the loss of ice by blasting some of it into space. It would be better to use lots of small asteroids. Think shot gun instead of cannon ball. Another possibility would be the production of greenhouse gases on Mars. Heat trapping fluorocarbons or chlorofluorocarbons could be manufactured and released into the atmosphere. It is known that there is chlorine in the soil of Mars and there's plenty of carbon in the atmosphere that could be extracted. Deposits of fluorine bearing minerals would have to be located.

If Mars could be warmed up enough to cause water ice to melt and vaporize, a thicker atmosphere of $CO_2$ and water vapor would form and trap more heat. The pressure could become high enough for liquid water to exist there. Algae could be planted and it could convert the $CO_2$ to oxygen. Higher plants and animals could be introduced and Mars could come to life. A whole new planet would be available for humanity and other life forms.

It's a beautiful dream and I don't know how realistic it is. The main thing is the creation of liquid water on Mars. Algae can double its mass four times a day. In a few years the whole planet would be crawling with algae that might rapidly transform the $CO_2$ to oxygen.

A terraforming project would be expensive and it might not pay off for a thousand years. Would an international program pay for it or would people with a passion for Mars spend all their money to live there, bootstrap, have children and populate Mars until there were enough workers and industry to build the greenhouse gas factories and associated industries including chlorine and fluorine mining?

The Moon and Earth orbital space can probably get everything needed from the Moon, near Earth asteroids and Earth. The Martians could get everything they need from local resources. They would make everything themselves and trade with one another. Why would they need to export anything to the Moon or Earth orbit? What would they do with Earthly money? Could they use terran money to buy high tech equipment like futuristic quantum computers and AI software? Or could they pay the spaceship owners to move more settlers to Mars? Sort of like Europeans coming to America and sending money back to the old country so their relatives could come to the USA.

Artists and musicians on Mars could be very special. They might attract lots of legal tender from the old world so that the Martians could buy things Earth has that they don't. Is there anything they could sell to the Moon or settlements in orbit? The Martians would have valuable carbon, but most of that could come from asteroids along with hydrogen and nitrogen. Maybe they would have uranium mined from rich deposits on Mars. Nuclear fuels could be used to propel spaceships throughout the solar system, presuming helium 3 fusion drives are somehow impractical. Lightweight vapor core fission reactors may be the key to the solar system. It's hard to envision interstellar travel without fusion, but ships with light sails or magnetic sails propelled by giant banks of lasers or particle beams energized by solar power plants in space hundreds of miles in diameter might be the way to reach the stars.

Unless there are unexpected ores on the Moon, there are two elements that Mars has that would be useful on the Moon-copper and chlorine. We know that there are perchlorates in the soil. These would have to be washed out before any greenhouse farming could be done in Martian soil. Chlorine is needed to make methyl chloride used in silicone polymer synthesis. It is needed to make salt when combined with sodium which the

Moon has. The Moon also has potassium which can be combined with chlorine to make potassium chloride which some people use as a salt substitute. Chlorine is needed to make PVC plastic and silane rocket fuel. It can be used to treat water, make bleach and dry cleaning fluid-carbon tetrachloride. With growing populations on the Moon and in free space settlements the demand for these substances may become worth supplying to earn money. Since it takes less energy to travel from Mars to the Moon than from Earth to the Moon, entrepreneurs might buy their chlorine from Mars, presuming that would be cheaper.

Martians could combine chlorine with copper to make solid copper chloride salts that are loaded into metal shells or basalt fiber bags and shot into space with mass drivers. The payloads could be captured in space and transported to the Moon and Earth orbit with robotic ships riding the solar wind with magnetic sails. Copper would be really useful. Mars has about 50 ppm copper in its regolith. The Moon only has about 14 ppm. It seems like Martians would have to mine and process millions of tons of regolith to get copper as would the Lunans. However, Nature may have concentrated copper in sulfide ores on Mars.[51] If this is true, then Mars could be an economical source of this valuable metal.

Copper is the preferred electrical conductor for all kinds of things even though aluminum is used for long distance power cables due to its lighter weight. Copper is used in generators, transformers, household wiring, telecommunication cables, electronics, heat sinks, heat exchangers, magnetrons in microwave ovens and other things. There are good reasons that the cables connected to your computer are made of copper instead of aluminum! It makes electric motors more efficient. Lunans should welcome copper to replace aluminum electric motor coils. On Earth, electric motors account for 43% to 46% of all electric power consumption and 69% of all the power used by industry.[52] Lunans and free space settlers will be just as reliant on electric motors if not more, since vehicles including trains and excavators will all be propelled by electric motors instead of internal combustion engines. Electric motors will be needed to drive water and sewage pumps, refrigerators and heavy cryogenic equipment, compressors, machine tools, hand power tools, ventilation and air filtration fans, rolling mills, hydraulic pumps for extruders and other machines, forced convection fans in weightless ovens, cooling fans, jackscrews for steering rocket motors, elevators, spin matching cars and centrifuges in spaceships, electric knives, kitchen equipment like mixers and blenders, conveyor belts, robots, cranes, etc. The electric motor may be as great an invention as the light bulb! Copper is also needed to make strong 2000 series aluminum alloys. It's even used for pigments, plumbing, architecture, jewelry and cookware. And it's antibacterial and antimicrobial.

Mars is the only planet besides Earth that might have ore bodies concentrated by hydrological processes. Mercury and Venus never had liquid water or life. Titan is too cold for liquid water. Europa and Enceladus may have subsurface seas of liquid water beneath their icy crusts, but we can't expect to find hydrothermal deposits of any kind and if we did they would not be easy to access. Mars might be of great financial value to solar system wide civilization after all!

# The Space Taxi

Let's consider a space taxi from LEO to GEO. If crews are going to be rocketed to LEO then travel up to GEO to work on SSPS to do jobs that can't be done by robots teleoperated by ground crews, there will have to be some kind of vehicle, a space taxi, to transport them. It's possible that the capsule the crews reach LEO in will have some kind of service module with rocket engines that is fully fueled for the LEO to GEO transfer when they reach orbit, or the service module will take on propellant at an orbital depot. There may have to be a depot in GEO to refuel. It's also possible for the capsule to dock with another vehicle, the space taxi, which is fueled in orbit, and ride that to GEO. Of course, things are simpler if the capsule to LEO is combined with a service module with rockets that can move them to GEO without rendezvous maneuvers, docking and transfer with the space taxi. Even so, if the service module doesn't have enough propellant to reach GEO at launch and it has to rendezvous with the depot, and perform docking and refueling procedures, efficiency is hampered. It's simpler to use a large enough reusable rocket to boost the capsule and service module with enough fuel to reach GEO in one piece.

What about travelers to Mars, if nuclear powered spaceships are not allowed any closer to Earth than GEO? There will have to be a similar system for them.

First, let's look at what is involved with a LEO to GEO transfer. The math for calculating the parameters of a minimum energy transfer, a Hohmann, between two orbits only requires algebra. In reality, things would get more complicated because of orbital perturbations by other bodies in the solar system, but it isn't too hard to get the basic numbers. In this age of the home computer, things are even simpler. Using the Quick Orbits program by Delta-Utec Space Research and Consultancy, 1999, we find that for transferring from LEO at an altitude of 500 km. to GEO at 36,000 km., which is a rounded off figure, the orbital period of the ellipse is 10.7 hours. Since the time of flight is half of that, it would take 5.35 hours or 5 hours and 21 minutes to make the flight. That's not too much time to stay in a cramped capsule for roughneck work crews. Hopefully, they will be lean and physically fit; not given to overeating and the side effects of gluttony.

They will eat a very light low residue dinner the night before and eat a high protein low residue breakfast consisting of steak and eggs, with some orange juice. If worse comes to worse on the 5 hour 21 minute flight in the capsule, there are always doggie bags.

This won't be acceptable to most tourists, accustomed to the good life, except maybe for the adventure tourist types who like mountain climbing, sky diving, SCUBA diving and base jumping. Mars settlers might be a hardier breed. It should be possible to modify a chemically propelled inter-lunar tourist ship, which is about as comfortable as a modern airliner, and move them to GEO in relative safety and comfort. The ships I envisioned for lunar tourism were about 27.5 feet in diameter and had two passenger decks. One had 80 passengers and one had 120 passengers. There are 4 toilets on each end of each deck for a total of 16 toilets for 200 people. There are floating rooms and sleeping cubicles in the rear hull that might be replaced with more seats and a few toilets. Floating rooms and sleeping cubicles won't really be needed on a roughly five and a half hour flight. This vessel would be more of a space bus or a space ferry boat than a space taxi!

Getting back to the roughnecks, how long do they stay up there? Isn't it going to be expensive for their employers to fly them up to GEO for a 12 hour shift and bring them back down every day? Yes. They will have to stay in space stations; perhaps inflatables tethered together by a super strong Kevlar cable that rotate to produce some healthy Gees. When needed, they can go space walking and unstick the robots or do jobs the robots cannot. High orbit is not within Earth's protective geomagnetic field and they won't get much cosmic ray blocking effect from Earth below. Astronauts on the ISS or people staying in the ELEO Kalpana stations don't endure as much cosmic ray exposure as our boys in GEO will.

By using OLTARIS, we find that 60 days at 36,000 km. high orbit with a 0.3 cm aluminum skin and 12 cm thick polyethylene layer for the habitat, will result in a total radiation exposure from cosmic rays, trapped protons and neutrons of 65 millisieverts or 20.5 milliGrays. The eyes and skin will receive a dose of 57.5 mSv and the blood forming organs will get 50.6 mSv. Compared to the radiation dose limits of terrestrial radiation workers this is a lot. The limits are 20 mSv/yr. averaged over five years with no more than 50 mSv in one year. The limits for the eyes are the same. For the skin, the dose can be 500 mSv in one year.[53] For comparison, a whole body CT scan will give a dose of 10 mSv.[54]

Astronauts can receive much higher doses. The career limits for males are 2000 mSv + 75 x (age in yrs. Minus 30) mSv. For a 40 year old male the limit would be 2750 mSv. The limit for the eye is 4000 mSv and for the skin 6000 mSv. The formula for females is 2000 mSv + 75 x (age minus 38) with the same career limits for eyes and skin. The 30 day limits are 250 mSv for the blood forming organs (BFO), 1000 mSv for the eyes and 1500 mSv for the skin. The annual limits are 500 mSv, 2000 mSv and 3000 mSv for the BFOs, eyes and skin respectively.[55]

Workers will endure more radiation exposure if they have to take space walks. It looks like they will have to endure more radiation than ground based workers, but they won't exceed the limits set for astronauts. Spaceship crews will endure more radiation. If they make two lunar round trips per week they are going to be exposed to lots of radiation over a period of a year. At 36,000 km. altitude almost 400 mSv are absorbed in years' time with the BFOs getting about 300 mSv and the eyes and skin 350 mSv. Lunar round trips will mean a similar amount of exposure; however, spaceship crews will spend some time after each 30 hour flight in well shielded space stations in LEO, LLO or at L1. This could give their bodies some time to repair DNA damage. Chances are the interlunar crews won't work in space for much more than a year, to be safe.

Mars settlers won't get dosed by a lot of radiation on their way to GEO but they will spend about 39 days in interplanetary space on their way to Mars. After that they will spend most of their time in underground habitations where they will be protected from cosmic rays and other dangers.

The problem of space radiation is a complex one. This is a very cursory treatment of the subject. Radiation dangers must not be scoffed at by those with cavalier attitudes. It can cause cancer, cataracts, tissue damage, nervous system degeneration and mutations. Space workers and settlers must not be smokers or heavy drinkers and it will be better to hire older persons beyond their child bearing years.

# ISRU Energy Storage

There isn't enough cadmium, lead or lithium on the Moon to make rechargeable batteries for storing energy during the two week long lunar nightspan. There is enough nickel, iron and potassium to make FeNi alkaline batteries like those invented by Edison. These batteries can withstand lots of abuse in the form of overcharging, over-discharging and short circuiting. Deep cycling does not reduce their life spans. These rugged batteries can last 30 to 50 years.[56]

There are meteoric iron and nickel particles in the lunar regolith in concentrations of a few tenths of a percent by mass. These could be scavenged by rovers with low intensity magnetic separators that dig through many square kilometers of lunar surface to a depth of about a meter. The particles are fused with silicates. They could be purified by screening, sieving, grinding and magnetic separation to get pure iron-nickel along with tiny amounts of other meteoric elements like cobalt, germanium, gallium and platinum group metals.[57] Treatment with CO gas to separate the iron and nickel is one way to get the nickel, but use of a supersonic dust roaster and all isotope separator would be even better. Such a device would let us get the cobalt, germanium, gallium and PGMs also.

Potassium is present in regolith at a concentration of a few tenths of a percent. It could be roasted out directly along with sodium and sulfur at 900 C. to 1200 C. It could also be obtained when processing regolith in SDR-AISs to get oxygen, silicon, aluminum, iron, titanium and other elements.

The potassium would simply be reacted with water to make an electrolyte solution of potassium hydroxide for the batteries. This reaction releases hydrogen which would be captured for other uses. The iron and nickel would be shaped into tubular or flat plate electrodes. The battery casing could simply be made of cast basalt which resists acid and alkaline attack.

The batteries would be kept inside habitat to protect them from temperature extremes. The habitat would be covered with several meters of regolith that serves as excellent thermal insulation. A fused regolith habitat module made with a large "3D printer" that is 20 ft. wide and 35 ft. long would have an internal volume of nearly 11,000 cubic feet not counting the domed ends. These batteries hold about 30 Watt hours per liter. A cubic foot is equal to about 28 liters. We will estimate that the batteries consume about 25 liters per cubic foot and only 10,000 cubic ft. of space in the module is used. That means there would be 250,000 liters of batteries that can hold 7,500 kilowatt hours of power. If the lunar night is 336 hours long, this bank of batteries could supply 22 kilowatts constantly during the whole time. This might not be enough to power a mass driver that can launch huge amounts of material into space every year, but it should be enough power to illuminate and power the life support systems of a small manned Moon base.

For comparison, the average US household in 2019 used 10,649 kW hrs. per year or about 877 kW hrs. per month.[58] So for two weeks this would be equal to about 8.5 times the power that is used by the average residential consumer in one month. The Moon base will have more than a family of four and crops will demand lots of power for nightspan illumination.

Numerous battery modules could be made to store more power. Cast basalt casings could easily be produced by machines. Small imported bench top rolling mills could be used to make the iron and nickel plates. These Edison batteries are very heavy (19-25 Whr./kg.) but that won't matter for stationary applications. Imported lightweight lithium batteries (250-700 Whr./kg.) used before mining and manufacturing are going on could be repurposed for vehicle, mining and construction machinery power.

Low mass foil reflectors will be landed early on. The reflectors could concentrate sunlight and increase the output of imported GaAs/Ge/GaInP multi-junction solar panels by a factor of ten or more. Half the power generated by day could be stored for night in lithium batteries. Lithium batteries are expensive but shipping by rocket costs even more; so lithium batteries will be landed on the Moon because of their light weight. They will also need protection from lunar thermal extremes and that will be provided by inflatable habitat covered with several meters of regolith. The combination of multi-junction solar panels and reflectors with lithium batteries will be very efficient, but there's no way to replicate them on the Moon given the available resources. Silicon solar panels and FeNi alkaline Edison batteries are "Moon makeable" and should be sufficient.

There may be other ways to store energy. Electric heat could be used to melt salts stored in insulated containers buried under a meter or so of regolith, since regolith is a good thermal insulator. The heat from the molten salts could be tapped directly to keep equipment warm in the brutal cold of the lunar night that could cause metals to crack. Another way to heat up the salts would be with solar furnaces that focus the Sun's rays through a quartz window in the salt container or by using trough reflectors and boiler tubes filled with molten salt. The only question is, what does the salt weigh compared to other energy storage systems like lithium batteries, flywheels, fuel cells and associated tanks, pipes, electrolysis units, and compressed gas storage systems? Perhaps heat could be stored in large blocks of cast basalt made by digging a ditch in the ground for a crude sand mold, melting some mare regolith in a ladle furnace, and pouring the stuff into the mold. If solidified basalt, much different from regolith consisting of sharp particles with vacuum in between them, is a decent heat storage medium like concrete used in solar thermal systems today, the question becomes, should it be heated electrically or with the direct application of solar energy?

The lunar night is a serious challenge. Mass drivers that launch materials to L2 must be located on the lunar equator at about 33.1 degrees East. This means that two weeks of hot Sun and two weeks of cryogenic temperatures must be endured, unlike the polar regions where solar energy is available 80 to 90% of the time. The hot Sun can be coped with by using foil reflectors as shields. The cold demands more. Vehicles, excavators and construction equipment could be parked on slabs of cast basalt that are heated up in the direct sunlight by day and covered with foil shields to keep heat from radiating away into space in the form of infrared radiation by night. Other pieces of equipment and tools could be stored for night in the same way. Even so, it seems there will be a greater need for warmth and energy to provide it.

A small nuclear power plant would be ideal. The SNAP-10A fission reactor was launched into Earth orbit in 1965. It had a thermoelectric power conversion system that generated 600 watts electrical, which isn't a lot for our purposes. However, it generated 30,000 watts thermal and that's what we really need to keep our metal machines from cracking in the lunar night. Getting a reactor to the Moon in today's political climate would be difficult.

Radioisotope thermoelectric generators (RTGs) have been launched. One of the most interesting RTGs is the GPHS-RTG used on the Cassini, New Horizons, Galileo and Ulysses spacecraft. It generated 300 W electrical and 4,400 W thermal.[59] Several of these may be just what is needed for a bootstrapping lunar mining base. The SNAP-10A had a mass of about 431 kg. and the GPHS-RTG was about 57 kg. The difference between a fission reactor like the SNAP and an RTG is that the reactor uses a chain reaction and can meltdown while the RTG uses natural decay heat. The RTG is less dangerous but we wouldn't want one to re-enter the atmosphere or be thrown into a populated area by a rocket explosion. An RTG cannot meltdown and it can make power for decades. Given the realities of lunar nightspan and the need to provide the Earth with abundant, clean, carbon free energy it seems that RTGs will be essential for a lunar mining base and SSPS building program. Perhaps rockets with RTGs onboard could be launched from remote locations like French Guiana for safety's sake.

It seems as if two weeks won't be long enough to land and deploy all solar power and storage systems with teleoperated robots and build some kind of heat storage system. Some RTGs could be the only realistic way to provide power for keeping equipment warm by night in the early stages of lunar industrialization. If the solar panels, lithium batteries, excavators etc. can't make it through the lunar night the whole project will fail. Besides RTGs there would have to be some kind of system of pipes filled with silicone oil perhaps to move heat to the equipment. Humanity's future in space and on Earth depends on solving these problems.

# Fire and Ice

## Worlds of Fire: Mercury and Venus

The smallest of the eight planets, Mercury, seems as inhospitable as Venus with its even hotter surface cloaked in clouds of carbon dioxide and sulfuric acid. It takes 59 days for Mercury to rotate once. Its temperature ranges from -280 F. at night to 800 F. in the equatorial zones. At the poles, it is always below -136 F.[60] The surface of Mercury is a challenge even for robots and some bulky super-insulated spacesuits would be needed for humans. Even the Moon is less hostile, and Mars is a picnic by comparison.

Even so, little Mercury has attractions. Solar energy is 6.5 times more intense on average (it varies due to the planet's eccentric orbit), so solar power plants would be capable of generating plenty of power for life support systems and mass drivers which could work on the airless surface. Superconducting mass driver coils might simply be protected from the big hot Sun of Mercury with foil or sheet metal solar shields. The ground might conduct heat from areas outside of the shield shadows, but things might not be so bad. If Mercury's surface consists of some kind of regolith powdered by eons of meteoric bombardment like the Moon's, it's probably a poor thermal conductor and the ground within the shadow might not get too hot. There may be a depth below the surface at which temperatures are bearable and stable. On the Moon, it is a constant -4 F. (-20 C.) just a meter or so deep at low latitudes despite the intense heat and cold thanks to the insulating qualities of lunar regolith. Habitations might be built underground on Mercury that are also protected from radiation including solar flares and micrometeorites.

The polar regions may be the best places to dig in and build bases. Water ice in permanently shadowed polar craters, as on the Moon, has been detected. It is thought that there are about 100 billion to a trillion metric tons of ice on Mercury and that could support plenty of humans with recycling. Depending on how fast it could be mined, hydrogen from the ice might be combined with silicon to make silane and oxygen could be extracted from the water and rocks.

The surface of Mercury has not been sampled, but the *MESSENGER* spacecraft detected calcium, helium, hydroxide, magnesium, oxygen, potassium, silicon, sodium and water vapor in the planet's tenuous atmosphere which is so thin it's barely there and wouldn't hamper mass drivers nor allow aerobraking.  These elements are probably vaporized by comets impacting the surface and solar winds blasting the surface.  Since the surface of the Moon and stony asteroids consist of iron, magnesium, calcium and aluminum silicates as well as minor amounts of other elements, chances are these are to be found on Mercury also. There is probably no shortage of silicon and oxygen in the rocks.

Mercury is smaller than Ganymede and Titan, but denser and more massive.  It is so dense and massive that the gravity on Mercury is about the same as Mars'.  There seems to be a large iron core composing about 55% of the planet's volume, surrounded by a mantle and crust of silicates.  There are lava flows from ancient low shield volcanoes.  Perhaps heavy metals from the molten core were carried up with these lavas.

Another interesting thing about Mercury, is that due to its short 88 day orbital period around the Sun, the planet has very brief synodic periods with all the other planets of the solar system.  That means travelers from or to Mercury don't have to wait so long for launch windows.  Travel to or from Venus is possible every 145 days, Earth 117 days, Mars 101 days, Jupiter 90 days, Saturn 89 days, Uranus and Neptune, every 88 days.  Peter Kokh has called Mercury the "Grand Central Station of the Solar System."  We can imagine starships entering our solar system with passengers planning to travel to many different worlds.  We can't expect some massive star ark to slow down to interplanetary speeds and drop passengers off one world at a time.  That might take longer than an interstellar journey.  They could get off at Mercury and board ships to Earth, Mars, Saturn or any other world of our solar system including free space settlements without waiting years and years for a launch window.  By locating a central meeting place on Mercury, diplomats from all over the solar system pressured for time might be able to schedule flights to and from there more efficiently.  It takes a lot of delta V to fly to or from Mercury which might seem like a barrier in this age of slow chemical rockets, but nuclear propulsion systems of the future could knock that one down.

Transportation on Mercury would probably all be underground.  Surface vehicles would need massive thermal insulation systems except for very brief sorties.  Mercury has a weak magnetic field but it probably can't do much to shield us from galactic cosmic rays and a solar flare blasting away at close range would probably be deadly unless the vehicle had tons of shielding for every square meter of the hull.  Robots would have to bore through the regolith and create solid tubular walls or cut and cover to make subway tunnels.  At least there wouldn't be any ground water to contend with.  Mining machines could be shadowed with lightweight foil or sheet metal shields by day and use waste heat from their onboard nuclear reactors to stay warm in the supercold of night.  Extreme cold can make metals brittle and crack.

Mercury has a spin-orbit resonance of 3:2.  That means it sees one day every two Mercurian years or 176 Earth days.  When the Sun shines there is plenty of solar energy.  During the prolonged night massive energy storage systems or nuclear power would be called for. Another possibility is the construction of a globe circling solar power grid at high latitudes near the shadowed craters and ice deposits. Silicon solar panels would probably degrade too rapidly in the high heat and radiation. Gallium, arsenic and other elements for multi-junction panels might be too rare.  Parabolic or trough reflectors of aluminum, magnesium or polished steel could focus solar energy on boiler tubes filled with molten sodium and potassium along with supercritical $CO_2$ turbines and generators.  The liquid sodium and potassium could be drained at night and stored for heat. The axial tilt of Mercury is so small that it will not cause seasonal variations, but the planet's orbit is highly elliptical so solar energy varies from 5 to 10 times as intense as at Earth's distance from the Sun. The planet is drenched with energy, but the machinery needed to tap that energy and the difficulty of working on the surface given the temperature extremes means it won't be so easy or cheap to tap that energy.  Robots will have to go to work on Mercury before humans in large numbers and prepare habitations, factories, farms, subways, mass drivers, power plants and grids to energize everything.  Terraforming Mercury will never be possible, but civilization could exist underground.

Venus is another world of extreme heat, but the thick atmosphere conducts plenty of heat from the day side to the night side, so it doesn't ever get supercold on Venus.  It gets hotter than Mercury, about 900 F., and the pressure of its almost entirely $CO_2$ atmosphere is about 90 times greater than Earth's at sea level.  The heat and the crushing pressure, not to mention the sulfuric acid clouds, makes Venus the most inhospitable planet with a solid surface in the solar system.  The outer planets don't have solid surfaces, just oceans of liquid gases.

The friendliest place on Venus is at high altitude, about 30 miles (50 km.) up in the atmosphere, where temperatures range from 86 F. to 176 F. [61] The acidic clouds make this location difficult too.  Perhaps robotic factories suspended by balloons filled with hydrogen could be stationed there.  They could pump down $CO_2$ and nitrogen, compress, cool and liquefy these gases, and rocket them into orbit with nuclear thermal rocket shuttles that use liquid $CO_2$ for working fluid.  Power could come from onboard nuclear reactors, since the clouds obscure the Sun, or they could dangle boiler tubes in the lower layers of atmosphere to tap the heat of Venus and run turbogenerators.  The sulfuric acid could corrode everything, unless the machinery was plated inside and out with iridium or platinum from metallic asteroids.  In this way the atmosphere of Venus could be mined for carbon, oxygen and nitrogen.  Perhaps asteroids could be diverted into orbit around Venus and they could be mined to get all the materials needed to mass produce these robotic balloon borne mining stations.  Some interesting re-entry vehicles and deployment systems would be required.

There isn't enough water in the clouds of Venus to support life.  We can't inject blue-green algae into the clouds and convert the $CO_2$ to oxygen. Perhaps in the distant future, centuries from today, there will be enough industry in space and AI machines to make it possible to mine Mercury, asteroids and the atmosphere of Venus and build a gigantic solar shield at the Sun-Venus Lagrange point one, about a million kilometers from Venus towards the Sun.  It could be more than just a shield.  It could consist of modular solar power plant units that generate power for interstellar laser or particle cannon propulsion beams that drive light sail or magnetic sail fitted spacecraft up to decent fractions of light speed.  Maybe enormous particle accelerators to make anti-matter for starships could be energized by the power plants.  This would be a mega-scale engineering project.  The shield might be hundreds of thousands if not millions of kilometers in diameter. It might be the largest structure humans ever build.

Who would pay for this? Perhaps the United Federation of Worlds and all its space faring member nations would contribute to this thousand year project with the promise of receiving a piece of land on Venus once it has been cooled down by intercepting solar energy with the shield. Sulfuric acid in the clouds might react with elements in the rocks once it is cool enough and form water and sulfate salts. Small seas might form that are amenable to natural or genetically engineered photosynthetic microbes and slowly convert the $CO_2$ to oxygen.

Even if a breathable atmosphere never forms, it might become possible to build domed cities and sealed subways where people could live and travel. Aviation would be possible with hydrogen powered jets that get their fuel from water and release the vapor back into the cooler atmosphere. Alternatively, $CO_2$ could be converted to CO and that could be used as fuel if there is enough oxygen in the atmosphere. Even if a breathable atmosphere for humans never forms, it might be possible to cultivate genetically engineered crops on the surface of Venus once it has been cooled. The biggest problem would be water. Venus is very dry. Comets might be diverted onto collision courses with the planet before any construction is done on the surface to add water and other substances like ammonia that could dissociate into nitrogen and hydrogen. Comet impacts might lead to explosive blow-off and thin the thick heat trapping atmosphere and kick up enough dust to cause a sort of "nuclear winter" on Venus and help cool it down. There are perhaps a trillion comets in the Oort Cloud that only need to be nudged to get them to fall towards the Sun. Perhaps settlers in Bernal Spheres surrounded by thick layers of ice for radiation shielding out in the Oort Cloud someday could be recruited for this project. How would they travel to the inner system? That could take decades even centuries at the speeds of nuclear ships. There isn't enough solar energy out there in the boondocks of the solar system to power propulsion beams. Perhaps the way to get around in the Oort Cloud is by the use of anti-matter powered ships. If they send Venus some comets, we could pay them with anti-matter. It would be wise to only divert comets that are at oblique angles towards Venus when they crash into it and not in line with the shield. Some comets are certain to miss their target.

Such projects might be undertaken only by people with extremely long life spans and even then they might require several generations to complete.

Out of control population growth would be prevented by mastery of human reproductive biology and waiting until one is say 150 years old to have kids. A trillion smiling babies and a trillion consumers with needs and wants won't make anyone happy if they all come too fast too soon and even robot armies can't keep up with the breeding rate. Instead of a space faring utopia, aside from some of the annoyances of corporeal existence, we could turn the galaxy into a cesspool if we don't breed responsibly. Family planning is a fact of life and will remain so forever, especially with life span extension. Malthusian nightmares can only be avoided by reproductive medical science and not by space colonization.

Experience gained with the terraforming of Mars and Venus could be applied to similar worlds orbiting nearby stars. The creation of habitable worlds and the spreading of life in a galaxy where life may be terribly rare could give people purpose. Otherwise living for centuries could be a curse rather than a blessing. It would certainly be better to discharge aggressive energies by engaging in mega-scale projects, diverting comets, exploring the solar system and even nearby star systems than fighting space wars with some evil empire or hostile aliens. All those AI robots and other machines including quantum computers are going to need some really sophisticated software so IT specialists will never run out of things to do. Of course, we might create AI that can program AI, but will this replace or augment human programmers?

## Ice Worlds: Outer Planet Moons

The moons of the solar system, except for the Moon of Earth, Io and maybe Titan, are all covered with thick layers of ice. There are 19 moons of the solar system, if we include Neptune's Proteus, that have enough gravity to mold them into rounded shapes. Ganymede and Titan are larger than Mercury. Seven moons are larger than the dwarf planets Pluto and Eris. If these worlds orbited the Sun they would be classified as planets or dwarf planets.[62]

There are over 200 moons in the solar system. Earth, of course, has one— The Moon. Mars has two small moons. Jupiter has 79, Saturn has 82, Uranus has 27 and Neptune has 14 known moons.[63] Some of these moons are several hundred to thousands of kilometers in diameter and spherical. Others are irregular chunks of rock and ice just a few kilometers wide or less. Dwarf planet Pluto has 5 moons.  There may be undiscovered moons out there.  Several dwarf planets have small moons.

These distant moons of the outer solar system have rock and metal cores covered with ice.  The ices consist of frozen water, methane and ammonia.  Europa and Enceladus are believed to have subsurface oceans.  At hundreds of degrees below zero the ices will be extremely hard.  Perhaps the ices can be mined with lasers or electron beams that can cut out chunks of ice. Another possibility is the use of explosives to blast out ice in pit mines.  Mass drivers energized by nuclear power plants, since solar energy in the outer solar system is so weak, could launch the ices into space for use by space settlement dwellers.  Water has many uses.  Methane can supply carbon and hydrogen. Ammonia can supply nitrogen.  Thick layers of ice surrounding metal settlement hulls could serve as cosmic ray shields.

Ice mining on the moons might compete with mining asteroids for water and hydrocarbons.  Metals from asteroids would still be in demand for space settlement construction.

Titan, the only moon with a thick atmosphere, is composed of rock and ice.  It is thought to have a crust of ice and a subsurface layer of ammonia rich water.  There are lakes of methane and ethane in the polar regions.  Perhaps there are other heavier hydrocarbons in these lakes.  The methane and ethane could be pumped up and used as propellant for nuclear thermal rockets.  It could also be partially oxidized with oxygen from water ice to make a mixture of hydrogen and carbon monoxide called synthesis gas.  This gas could be reacted with the right proportions of CO and hydrogen, at the right temperatures and pressures with the right catalysts, to make almost any organic chemical for plastics, synthetic fibers, paint, drugs, solvents, elastomers, adhesives, etc. In 2008, scientists said that the lakes of Titan had hundreds of times more liquid hydrocarbons  in them than all the known natural gas and oil reserves on Earth. [64]

We can imagine bases on Titan where inflatable habitat is made from locally sourced organic chemicals.  Hopefully, rock outcroppings could be found where metals could be extracted and glass could be made.  Refineries could be built where chemicals are synthesized and shipped out to space settlements with huge nuclear tanker rockets.  Mass drivers can't be used due to the thick atmosphere, but we can envision mag-lev launch tracks where ships with nuclear ramjets are launched.  Due to its distance from the Sun and the haziness of the atmosphere, Titan's surface receives only 0.1% as much light as does the Earth.[65]  Solar power seems to be out of the question, unless we build SSPSs with vast reflectors to collect the dim light that far away in the solar system and beam it down to the surface.  Wind power might work on Titan.  Nuclear power would probably be the most practical.

Pumping up liquid hydrocarbons and refining them in mostly automated chemical plants could be so productive that Titan becomes a better and less expensive source of synthetic materials and organics than the ice mines of various moons and carbonaceous chondrite asteroids.  The inner solar system might have plenty of carbon from Venus and hydrogen from asteroids while Titan becomes the main source of these light elements and substances derived from them in the outer solar system.

Robotic supertankers and freighters might haul materials from Titan all the way down to the Main Belt and out as far as Neptune.  Even with an abundance of organics from Titan space settlers will have to reuse, repurpose and recycle anything made from organic chemicals.

Beyond the moons of the giant planets, there are many ice covered bodies in the Kuiper Belt and Oort Cloud that are rich in hydrogen, oxygen, carbon and nitrogen.  If metals are in short supply out there it might be possible to build free space settlements from synthetic materials like carbon fiber reinforced composites.  Ice would make an excellent radiation shield.  Harvesting sunlight at great distances would require huge reflectors.  It might be possible to run fusion reactors with deuterium or even protium from water, methane and ammonia ices.  Life in this region of the solar system presents many challenges, but space settlers might someday accept those challenges.  The resources out there beyond Neptune are free for the taking and there probably won't be much competition since only the hardiest pioneers would want to settle the boondocks of the Kuiper Belt and Oort Cloud. Eventually they might even reach the Oort Cloud of the Alpha Centauri A, B and Proxima star system.

# Dwarfs and Giants

**Dwarf Planets**

According to the IAU, the International Astronomical Union, there are five dwarf planets—Ceres, Pluto, Eris, Haumea and Makemake. These bodies are roughly spherical or ellipsoid and believed to be in hydrostatic equilibrium. This means that their gravity is countered by a pressure gradient force that causes them to be nearly round.[66] They range in size from the smallest, Ceres at about 945 km. diameter to the largest, former planet Pluto, at about 2375 km. Eris is a close second at 2325 km. There may be thousands of other dwarfs out beyond Neptune yet to be discovered.[67]

There are numerous other bodies in the solar system that would qualify as dwarf planets in my estimation, but I don't want to argue with the IAU. They aren't exactly round. They could only be called ellipsoid or irregular and some probably don't have melted and differentiated interiors. However, they are pretty big and would make interesting destinations for exploration. Ceres is classified as a dwarf planet and not just the largest asteroid. It is about 27% the diameter of Earth's Moon and it seems to have an icy mantle surrounding a rocky core. Vesta, the second largest asteroid, is about 525 km. in diameter or 15% as wide as the Moon. It is thought to have a metallic iron-nickel core, an olivine mantle and a basaltic crust. Next is Pallas, at 512 km. or about 15% the diameter of the Moon, and it is believed to have a composition similar to Ceres. Hygiea, about 435 km. diameter or 12% of the Moon, is thought to be undifferentiated and is similar to a carbonaceous chondrite.[68]

Hygiea is not exactly spherical but it is somewhat more rounded than Vesta and Pallas which are more ellipsoid. It hasn't melted and differentiated and that's probably the case with smaller bodies. Interamnia, the 5th largest, about 350 km. wide, on down to Eunomia at 255 km., the 10th largest asteroid, are substantial bodies with masses of quadrillions of metric tons.[69] Billions of Island 3 sized O'Neill cylinders could be built of these, but that would be a crime against Nature. These worlds would be interesting to explore, mine judiciously and inhabit in domed and underground cities.

Large centrifuges could be built where people go to get a healthy dose of Gee forces. To me, these are worlds that although they are not spherical, are big enough to be called dwarf planets, but there has to be a cut-off point and the IAU has chosen hydrostatic equilibrium to be the main criterion, although there is some debate regarding that for ice worlds. In any case, interesting cultures could evolve in settlements on and within these bodies. Seeing the Sun from many different vantage points in the solar system would be so interesting, and the small yellow disk in the sky of a Main Belt asteroid must be a fascinating sight to see.

There are at least a million asteroids over 1 km. in diameter and innumerable small ones just a meter or more wide that could either be regarded as sources of raw materials or just navigational hazards. The 12 largest asteroids contain about 60% of the mass of the entire Main Belt with Ceres alone accounting for about half of that. The other 40% in the form of smaller bodies and boulders contains quadrillions of tons of material for building free space settlements and most of them would not be missed if harvested. I wouldn't be surprised if some wealthy individual or group of people laid claim to a small asteroid, just a kilometer or more in diameter, and built a private Xanadu on one of these islands in the sky. There may be an equal number of Trojan asteroids at Sun-Jupiter Lagrange regions that could be mined and converted to rotating settlements, and some of those might be claimed by people who want to create their own private bastions against the mundanity of greater society. Then there are the Centaur asteroids orbiting between Saturn and Uranus that would be far out places to inhabit and maybe build a private observatory closer to the stars. Besides astronomers and non-conformists, many poets and painters would want to live out there somewhere, or at least just visit.

## The Giants: Jupiter, Saturn, Uranus and Neptune

Jupiter and Saturn are classified as Gas Giants. They have no real solid surfaces and are composed primarily of hydrogen and helium like the Sun. It is believed that they have molten cores of rock and metal surrounded by mantles of liquid metallic hydrogen. Surrounding that are layers of molecular hydrogen and atmospheres of hydrogen and helium with clouds of water and ammonia. [70]

Uranus and Neptune are called Ice Giants.  They contain hydrogen and helium in their atmospheres but they are mostly carbon, oxygen, nitrogen and sulfur in the form of water, methane, ammonia and sulfides.  The presence of methane in their atmospheres gives them their bluish and greenish colors.  It's thought that they have small rock and metal cores surrounded by ice and supercritical water oceans with atmospheres of hydrogen, helium and methane mostly.  They do not contain metallic hydrogen like the Gas Giants.[71]

It seems that these Giant Planets could supply hydrogen and helium for huge solar system wide populations forever with recycling.  They could even provide reaction mass for nuclear electric or fusion rockets for millions of years.  As for the development of some kind of reactionless drive that uses some kind of magnetic impeller that churns its way through the magnetic fields of the Sun or planets, that's borderline science fiction.  Carbon, nitrogen, oxygen and sulfur could also be mined by balloon borne factories floating in their atmospheres.  Jupiter's atmosphere is very turbulent and its gravity high with an escape velocity of 37 miles per second.[72]  Saturn is less stormy but it also has a high escape velocity of 22 miles per second.[73] Uranus and Neptune are farther away and colder, but they have less gravity and lower orbital and escape velocities.  It may be much easier to mine their atmospheres for hydrogen, helium, carbon, nitrogen, oxygen and sulfur and rocket tanks of these elements into orbit.

# Diet for a Small Solar System

In space, space is at a premium. Collections of modules, underground communities, domed cities and free space settlements will all have a limited amount of space for farming. An Island 3 O'Neill cylinder 4 miles wide and 20 miles long would have about 250 square miles of area within. Several square miles could be forested with fruit, chestnut and walnut trees that supply food as well as habitat for deer, rabbits and squirrels; birds too. There could be ponds stocked with fish, frogs and turtles. Youngsters camping in the forests could listen to stories about how their ancestors lived in forests on Earth and foraged to survive. There could be vertical farms 500 to 1000 meters high with levels one or two meters high for crops. If the vertical farm covers an area one acre at the base it would contain 250 to 1000 acres of agricultural space. There could be numerous vertical farms in Island 3 type settlements. Not all space settlements will be this large.

In habitat of all kind, temperature and light levels will be controlled. There will be no winter so three even four harvests might be brought in every year. Even greater efficiency could be achieved if space settlers consume a mostly plant based diet. In the USA, half the farmland and 78% of the grain is used for livestock feed. It takes about 21 pounds of plant protein to make one pound of beef protein. Milk and eggs require about 4.4 and 4.3 pounds respectively of plant protein for every pound of protein for human consumption. It seems it would make more sense to eat plant protein instead of so much animal protein. If the discarded stems and leaves of crops are fermented and used to feed dairy cows or other ruminants like sheep we could still have milk, cheese, butter, cream and yogurt without much waste of space.

Plant protein doesn't always contain all nine essential amino acids. There are a number of combinations that do supply all of them. One and a half cups of beans or peas and four cups of rice have as much protein as 19 ounces of steak. Other complimentary combinations besides rice and legumes are: [74]

1) rice and soy

2) rice, wheat and soy

3) rice and yeast

4) rice and sesame seeds

5) rice and milk

6) wheat with milk or cheese

7) wheat and beans

8) whole wheat and soy

9) wheat, soy and sesame

10) cornmeal and beans

11) cornmeal, soy and milk

12) beans and milk

13) beans and sesame seeds

14) soy, wheat, rice and peanuts

15) soy, sesame seed and peanuts

16) peanuts and sunflower seeds

17) peanuts, milk and wheat

18) sesame seeds and milk

19) potatoes and milk

Two more foods of interest are:

1) Quinoa is grown for its seeds which contain all nine essential amino acids and plenty of vitamins, minerals and dietary fiber. It is also gluten free.[75] Uneaten parts of the plant could be used for silage.

2) Algae which is rich in protein, vitamins, minerals, carbohydrates, cellulose and oil. Algae can be cultivated in tanks illuminated by sunlight during the two week long day on the Moon. At night the algae could be harvested and used for livestock feed and even food for humans. Algae,

also known as seaweed, can be fed directly to fish like Tilapia. It could be fed as is to chickens and other animals as a supplement to fermented stems and leaves (silage) from unused parts of crops, or its oil could be extracted first and the high protein remnants could be used as feed. We must wonder if dried algae with or without its oil can be used to feed edible insects like grasshoppers, mealworms, termites and ants which can supply cheap protein and be used as chicken feed too.

Large tanks of algae could replace farm space that would be needed for livestock feed and people food and allow more room for humans or more space for crops. Yeast, a fungus, which doesn't require sunlight, could be cultivated in tanks for livestock feed and human consumption. Like algae, nutritional yeast is rich in protein, vitamins and minerals. It would be dried and heated to inactivate it before ingestion. A source of carbohydrate would be needed for the yeast. Algae might be that source. Mushrooms, also fungus, grown in stacks of trays, vertical farms, could also be raised in the dark and provide food for animals and humans. These fungi might be fed to edible insects also.

Lunans have to deal with the monthly day/night cycle so algae, yeast and mushrooms could be very important parts of the food supply on the Moon. Martians are lucky enough to have a 24.5 hour day/night cycle and they don't need to store energy for two weeks of darkness or grow some masses of seaweed and fungi. Even so, Martians will need to heat their plastic domes and farm space will be at a premium. They could build underground habitations and pipe sunlight within through fiber optic cables and systems of lenses and mirrors. It might require less energy to heat underground dwellings and farms, but there will be limits to how much habitable space they can create. Martians might also cultivate algae, nutritional yeast and mushrooms since these can be very productive with a comparatively small amount of volume. They might want to ferment some silage and feed a mixture of silage, algae, yeast and mushrooms to cows and sheep to get milk and other dairy products. This feed mixture and insects might be fed to chickens for eggs too.

# References

1) Brandon Gaille. 16 Satellite Industry Statistics, Trends and Analysis. 2019. https://brandongaille.com/16-satellite-industry-statistics-trends-analysis/

2) Space Based Solar Power. https://en.wikipedia.org/wiki/Space-based_solar_power#Advantages_and_disadvantages

3) Space Weather. https://web.archive.org/web/20110929152905/http://www.solarstorms.org/Svulnerability.html

4) Gallium Arsenide. https://en.wikipedia.org/wiki/Gallium_arsenide

5) David A. Dietzler. Mining the Moon: Bootstrapping Space Industry. Chapter 9. Lunar Tourism. 2020.

6) Tom Marotta and Al Globus. The High Frontier: An Easier Way. 2018.

7) Space-based Solar Power. https://en.wikipedia.org/wiki/Space-based_solar_power

8) T.A. Heppenheimer. Colonies in Space. Chp. 3 Power from Space. 1976. https://space.nss.org/colonies-in-space-chapter-3-power-from-space/

9) T.A. Heppenheimer. Toward Distant Suns. Chp.6 Large Space Structures. 1979. https://space.nss.org/toward-distant-suns-by-t-a-heppenheimer/

10) Space Fabrication Demonstration System. Pg. 3-5 table 3-1. https://ntrs.nasa.gov/api/citations/19790021042/downloads/19790021042.pdf

11) Advanced Automation for Space Missions. Chp. 4.2.2 Table 4.16. https://en.wikisource.org/wiki/Advanced_Automation_for_Space_Missions/Chapter_4.2.2

12) Bulent Yusuf. Fully Printed 3D Motor is World First. 2018. https://all3dp.com/3d-printed-electric-motor/

13) David A. Dietzler. Mining the Moon: Bootstrapping Space Industry. Chapter 2. Lunar Money Matters. 2020.

14) T.A. Heppenheimer. Colonies in Space. Chp.6-The Moon Miners. https://space.nss.org/colonies-in-space-chapter-6-the-moon-miners/

15) Tom Marotta and Al Globus. The High Frontier: An Easier Way. Pg.65. 2018.

16) Yuen Yiu. How Far Can Laser Light Travel? American Institute of Physics. 2018. https://www.insidescience.org/news/how-far-can-laser-light-travel

17) https://newatlas.com/nasa-tests-space-gps-pulsars-sextant/52961/

18) MIT Technology Review. An Interplanetary GPS Using Pulsar Signals. https://www.technologyreview.com/2013/05/23/178344/an-interplanetary-gps-using-pulsar-signals/

19) https://optocrypto.com/nasa-tests-celestial-gps-x-ray-positioning/

20) X.P. Deng et. al. Interplanetary Spacecraft Navigation Using Pulsars. National Space Science Center, Chinese Academy of Sciences, Beijing, China. https://arxiv.org/pdf/1307.5375v1.pdf

21) The Engineering Toolbox. https://www.engineeringtoolbox.com/stress-rotation-disc-ring-body-d_1752.

22) Synthetic Aperture Radar. https://en.wikipedia.org/wiki/Synthetic-aperture_radar

23) Food for Spaceflight. https://spaceflight.nasa.gov/shuttle/reference/factsheets/food.html

24) Top 10 Best Space Food. https://www.seeker.com/top-10-best-space-food-1765022320.html

25) Bryce L. Meyer. Achieving Earth Independence: How Food Will be Grown? Pg. 14. 2019. https://www.combat-fishing.com/animationspace/AEIBrycesFoodISDC2019RED.pdf

26) Electrodynamic Tether. https://en.wikipedia.org/wiki/Electrodynamic_tether

27) Mag-lev trains: Why aren't we gliding home on hovering carriages? The Guardian. 2018. https://www.theguardian.com/technology/2018/may/29/maglev-magnetic-levitation-domestic-travel

28) https://www.sciaky.com/additive-manufacturing/electron-beam-additive-manufacturing-technology

29) Rail Profile. https://en.wikipedia.org/wiki/Rail_profile#Europe

30) Bryce L. Meyer. Achieving Earth Independence: How Food Will be Grown? Pg. 14. 2019. https://www.combat-fishing.com/animationspace/AEIBrycesFoodISDC2019RED.pdf

31) https://boltthreads.com/

32) Rayon. https://en.wikipedia.org/wiki/Rayon#Manufacture

33) Ethylene Glycol. https://en.wikipedia.org/wiki/Ethylene_glycol

34) https://en.wikipedia.org/wiki/List_of_countries_by_the_number_of_millionaires

35) WWW.SULFURLAMP.COM General Technical Information page, Innovative Lighting: 2000. http://www.sulfurlamp.com/tech.htm

36) International Dark-Sky Association-Information Sheet 114. "Illumination Levels, Then and Now." excerpts fr. A Text-Book of Physics, Louis B. Spinney. http://www.darksky.org/~ida/infoshts/is114.html

37) "Table of Illumination Requirements." http://www.olemiss.edu/depts/environmental_safety/light.pdf

38) David H. Trinklein, "Lighting Indoor Houseplants." Agriculture publication G6515. University of Missouri, Columbia: 1999. http://muextension.missouri.ed/xplor/agguides/host/g06515.htm

39) Clemson University Cooperative Extension Service. Home&Garden Information Center. "Indoor Plants-Cleaning, Fertilizing, Containers and Light Requirements." HGIC 1450. http://hgic.clemson.edu/factsheets/HGIC1450.htm

40) Growco Indoor Garden Supply. Intense Lighting Tutorial from 4hydroponics.com. "High Intensity Indoor Lighting lets you outsmart nature." http://www.4hydroponics.com/lighting/lighting_help.htm

41) Mushroom Council. "Six Steps to Mushroom Farming." The Penn. State U. College of Agriculture, Extension Service. University Park, PA. http://www.mushroomcouncil.com/grow/sixsteps.html

42) https://www.epa.gov/indoor-air-quality-iaq/what-hepa-filter-1

43) https://alpinefreshair.com/hepa/

44) https://en.wikipedia.org/wiki/Ore

45) https://en.wikipedia.org/wiki/Aristarchus_(crater)

46) https://www.nasa.gov/topics/moonmars/features/moonrock-king.html

47) https://en.wikipedia.org/wiki/Oxyliquit

48) https://printermaterials.com/recycle-3d-filament-3d-printer-filament-recycler/

49) https://www.relativityspace.com/

50) Carl Sagan. Cosmos. Chp. 5. Blues for a Red Planet. Pg. 132. Random House, New York: 1980.

51) Robert Zubrin. The Case for Mars. Chp. 7. Building the Base on Mars. Pgs. 204-205. Touchstone, New York: 1997.

52) Copper. https://en.wikipedia.org/wiki/Copper#Applications

53) radiopedia.org/radiation/dose-limits

54) epa.gov/radiation/radiation-sources-and-doses

55) https://www.windows2universe.org/?page=/spaceweather/astronaut_dose.html

56) Nickel-Iron Battery. https://en.wikipedia.org/wiki/Nickel-iron_battery

57) Dr. William Agosto. "Lunar Benefication." http://www.nss.org/settlement/nasa/spaceresvol3/lunarben1b.htm

58) FAQs. https://www.eia.gov/tools/faqs/faq.php?id=97&t=3

59) Radioisotope Thermoelectric Generator. https://en.wikipedia.org/wiki/Radioisotope_thermoelectric_generator

60) Mercury(planet). https://en.wikipedia.org/wiki/Mercury_(planet)

61) Venus. https://en.wikipedia.org/wiki/Venus#Atmosphere_and_climate

62) https://www.planetary.org/space-images/the-solar-systems-major-moons

63) Moons. https://solarsystem.nasa.gov/moons/in-depth/

64) Lakes of Titan. https://en.wikipedia.org/wiki/Lakes_of_Titan

65) Titan. https://en.wikipedia.org/wiki/Titan_(moon)

66) Hydrostatic Equilibrium. https://en.wikipedia.org/wiki/Hydrostatic_equilibrium

67) List of Possible Dwarf Planets. https://en.wikipedia.org/wiki/List_of_possible_dwarf_planets

68) Asteroid. https://en.wikipedia.org/wiki/Asteroid

69) 10 Largest Asteroids Ever Detected (By Diameter). https://largest.org/nature/asteroids/

70) Gas Giant. https://en.wikipedia.org/wiki/Gas_giant

71) Ice Giant. https://en.wikipedia.org/wiki/Ice_giant

72) Jupiter. https://en.wikipedia.org/wiki/Jupiter

73) Saturn. https://en.wikipedia.org/wiki/Saturn

74) Frances Moore Lappe. Diet for a Small Planet. Friends of the Earth/Ballantine Books. New York. 1971.

75) Quinoa. https://en.wikipedia.org/wiki/Quinoa

Made in the USA
Monee, IL
07 February 2021